Application of Microbe-mineral Interaction for Remediation of Soil and Groundwater

Wang Wenbing

上海大学出版社
· 上 海 ·

图书在版编目（CIP）数据

微生物-矿物交互在土壤和地下水修复中的应用 = Application of Microbe-mineral Interaction for Remediation of Soil and Groundwater ： 英文 / 王文兵著. -- 上海 ： 上海大学出版社，2024. 12. -- ISBN 978-7-5671-5082-9

Ⅰ. X523； X53

中国国家版本馆 CIP 数据核字第 2024LB9990 号

责任编辑 司淑娴
封面设计 倪天辰
技术编辑 金 鑫 钱宇坤

Application of Microbe-mineral Interaction for Remediation of Soil and Groundwater

Wang Wenbing

上海大学出版社出版发行
（上海市上大路 99 号 邮政编码 200444）
（https://www.shupress.cn 发行热线 021－66135112）
出版人 余 洋

*

南京展望文化发展有限公司排版
江苏凤凰数码印务有限公司印刷 各地新华书店经销
开本 710mm×1000mm 1/16 印张 14.75 字数 250 千
2024 年 12 月第 1 版 2024 年 12 月第 1 次印刷
ISBN 978－7－5671－5082－9/X·15 定价 60.00 元

前 言

重金属、类金属、卤代有机污染物及新污染物等是国内外场地土壤和地下水中最常见的污染物，尤其是新污染物，其危害日益凸显和被广泛关注，而在当前国家“双碳战略”需要下，绿色低碳成为场地修复技术研发的主要目标和发展方向；微生物、零价铁及其他铁硫、铁氧矿物作为环保绿色低碳的材料也得到了广泛应用。微生物-矿物高效交互技术作为一种极具潜力的绿色低碳修复技术非常满足当前国家战略发展要求，这也对修复材料和技术在场地实际应用提出了更高要求。本书是作者多年绿色低碳修复材料研发科研成果和部分场地修复实际案例的总结，涉及的技术和理论对场地修复和应用、人才培养、国民经济可持续发展、低碳发展具有科学指导和实际应用价值。

本书从微生物-矿物的高效交互反应出发，从国内外视角全面和系统地介绍了微生物-矿物交互修复技术用于处置各类污染物的当前现状和发展；全书共分 7 章：第 1 章土壤和地下水污染现状及微生物-矿物交互基本概念；第 2 章当前微生物-矿物交互技术现状；第 3 章微生物-矿物交互技术用于处置重金属（六价铬（Cr(VI)）、类金属污染（砷(As)）和其他放射性物质（铀(Uranium)）污染；第 4 章微生物-矿物交互技术用于去除有机污染物（芳香烃(aromatic hydrocarbons)、氯代烃(chlorinated hydrocarbons)）；第 5 章微生物-矿物交互技术用于去除新污染物（抗生素(antibiotics)、全氟化合物(per- and polyfluoroalkyl substances)、溴代阻燃剂(brominated flame retardants)等）；第 6 章微生物-矿物交互在环境介质中涉及的多过程耦合反应模型；第 7 章其他物理-化学手段与零价铁耦合交互技术。

本书参考和引用前人大量研究成果，经过作者的理解、归纳和总结，给出微生物-矿物高效交互技术针对不同污染物的技术要点、机制差异和相关物理-化学-生物多过程耦合反应模型,同时系统提供微生物-矿物交互修复污染土壤和地下水的研究案例与部分应用实例。作者衷心感谢在这一领域做出贡献的学者和同行,没有他们的研究和长期积累,很难形成这一理论和科学体系,也衷心感谢伦敦大学学院(UCL)刘一鸣和上海大学范淇峰、张梦、龚天添、董纤凌、魏永康为各部分提出的建设性意见及对本书出版工作的帮助和支持,使得这部理论著作得以顺利完成。

同时感谢国家重点研发计划项目“场地地下水卤代烃污染修复材料与技术”(项目编号: 2020YFC1808200)和国家自然科学基金项目(项目编号: 42177386)的资助。

本书在编写过程中错误在所难免,恳切希望读者批评指正。

王文兵

2024 年 4 月于上海

Preface

Heavy metals, metalloids, halogenated organic pollutants, and emerging contaminants are the most common pollutants in soil and groundwater at domestic and international sites, especially emerging contaminants, whose hazards are increasingly prominent and widely concerned. Under the current national "dual carbon" strategy, green and low-carbon technologies have become the main goals and development directions for site remediation technology research and development. Microorganisms, zero-valent iron, and other iron sulfide and iron oxide minerals, as environmentally friendly and low-carbon materials, have been widely applied. The efficient interaction technology between microorganisms and minerals, as a promising green and low-carbon remediation technology, meets the current national strategic development requirements. Therefore, higher demands have been proposed for the practical application of remediation materials and technologies on site. This book is a summary of the author's many years of research and development of green and low-carbon remediation materials and some practical cases of site remediation. The technologies and theories covered have scientific guidance and practical application value for site remediation and application, talent cultivation, and sustainable national economy and low-carbon development.

Starting from the efficient interaction between microorganisms and minerals, this book comprehensively and systematically introduces the current status and development of the application of microorganism-mineral interaction remediation technology for various pollutant disposal from domestic and international perspectives. The book is divided into 7 chapters. Chapter 1 discusses the current status of soil and groundwater pollution and the basic concepts of microorganism-mineral interaction. Chapter 2 covers the current status of microorganism-mineral interaction technology. Chapter 3 focuses on the application of microorganism-mineral interaction technology for disposing heavy metals (hexavalent chromium), metalloids (arsenic), and other radioactive substances (uranium) pollution. Chapter 4 discusses the application of microorganism-mineral interaction technology for removing organic pollutants (aromatic hydrocarbons, chlorinated hydrocarbons). Chapter 5 explores the application of microorganism-mineral interaction technology for removing emerging contaminants (antibiotics, per- and polyfluoroalkyl substances, brominated flame retardants, et al.). Chapter 6 delves into multi-process coupled reaction models involving microorganism-mineral interactions in environmental media. Chapter 7 discusses other physical-chemical methods and Fe^0-coupled interaction technologies.

This book references and cites a large number of previous research results. Through the author's understanding, summarization, and deduction, the book presents the technical points, mechanism differences, and related physical-chemical-biological multi-process coupled reaction models of microorganism-mineral efficient interaction technology for different pollutants. It also systematically provides research cases and some application examples of microorganism-mineral interaction for remediating polluted soil and groundwater. The author sincerely thanks the scholars and colleagues who have contributed to this field. Without their research and long-term exploration, it would be difficult to form this theoretical and scientific system. The author also expresses gratitude to Liu Yiming from University College London (UCL) and Fan Qifeng, Zhang Meng, Gong Tiantian, Dong Qianling, and Wei Yongkang from Shanghai University for their constructive suggestions and support for the publication of this book, which enabled the completion of this theoretical work.

Acknowledgments to the financial support from the National Key Research and Development Program project "Remediation Materials and Technologies for Groundwater Contaminated with Halogenated Hydrocarbons" (Project No. 2020YFC1808200) and the National Natural Science Foundation of China (Project No. 42177386).

Errors are inevitable in the writing process of this book, and the author earnestly hopes for readers' suggestions and corrections.

Wang Wenbing

April 2024 in Shanghai

Content

Chapter 1
Introduction

1.1 Current situation of soil and groundwater contamination

With the progress of industrialization and urbanization, people pay more attention to soil and groundwater contamination (Abd-Elaty et al., 2020). Heavy metals (e.g. chromium (Cr)) and organic compounds (e.g. dyes and pharmaceutical compounds) are harmful to the ecological environment (Anandan et al., 2020). Human activities, such as mining, farming and manufacturing, are the main ways for contaminants to enter the soil and groundwater (Economou-Eliopoulos and Megremi, 2021). Moreover, groundwater accounts for about 30% of freshwater resources on the earth (Kalhor et al., 2019). The World Health Organization (WHO) calculated that in 2012, about 8.9 million people died due to contaminated soil, water, and air (Dermatas, 2017). The protection of groundwater has been given priority by the European Commission, EU Member States, and European national institutions (Hartmann et al., 2018; Lapworth et al., 2019). In addition, groundwater management policies such as the European Water Framework Directive, the Water Law in Israel and the Water Law in China have also been formulated (Zhang et al., 2019). Regarding the role of groundwater and soil, a convenient and practical method needs to be proposed to treat contaminated groundwater and soil. Microbe-mineral interaction technology such as the permeable reactive barriers (PRBs) with ZVI or mineral fillers installed underground were used to in situ eliminate contamination (Chen et al., 2011; Do et al., 2011). Furthermore, some indigenous microorganisms can form in the PRBs to accelerate contaminant attenuation (Wang et al., 2022).

1.2 Microbe-mineral interaction

Compared to artificial bioaugmentation or biostimulation, the natural attenuation effect of indigenous microorganisms produced in PRBs with ZVI or mineral fillers which were installed in underground was poor (Huang et al., 2021). The Bio-augmented permeable reactive barriers (Bio-PRBs) is a typical microbe-mineral interaction system. It focused on artificial bioaugmentation or biostimulation, which was referenced in the definition of Xin (Xin et al., 2013). Microbe-mineral interaction Bio-PRBs is a new type of PRBs that is constructed by microorganism activation or immobilized microorganism with other media (iron-based materials, biochar, etc.) (Motlagh et al., 2020).

Immobilized microorganisms have an excellent ability to adapt the environmental changes (Yang et al., 2022). And they showed great potential for remediating trichloroethene (TCE), petroleum, Cr(Ⅵ), and personal care products contaminated groundwater (Motlagh et al., 2020; Zaheer et al., 2021). The roles of microorganisms in microbe-mineral interaction Bio-PRBs include providing e^{-1} donors (Sathishkumar et al., 2016), facilitating e^{-1}transfer (Zhou et al., 2022), and degrading contaminants (Schostag et al., 2022). Currently, Bio-PRBs can be divided into two types: non-removable reaction barriers and removable reaction columns (Wang et al., 2022; Wang and Wu, 2019a). Compared to non-removable reaction barriers, the removable reaction columns are more conducive to installation, disassembly, and regular replacement of fillers (Li et al., 2021; Wang et al., 2022).

The fillers include microorganism activation materials, microorganism immobilization materials (carriers), and abiotic reactive materials. Glucose (Vijayanandan et al., 2018), wheat straw (Liu et al., 2021), glycerin (Kumar et al., 2016), compost leaf mulch (Angai et al., 2022), CaO_2 nanoparticles (Mosmeri et al., 2017), and coconut shell biochar (Liu et al., 2019) were the microorganism activation materials. They can promote the activity of native or added microorganisms for promoting the removal of *o*-nitrochlorobenzene (*o*-NCB) (Liu et al., 2021), As(V) (Angai et al., 2022; Kumar et al., 2016), Cr(VI) (Cancelo-González et al., 2015), Se (VI) (Sasaki et al., 2008), phenanthrene (PHE) (Liu et al., 2019) (Table 1.1).

Moreover, the immobilization carriers, peanut shell biochar (Wang et al., 2022), clay composite adsorbent (Vijayanandan et al., 2018), high-density luffa sponge (Wang and Wu, 2017), zero-valent iron (ZVI) (Yang et al., 2018), activated carbon (Huang et al., 2021), herbal fruit residue biochar and spruce biochar (Siggins et al., 2021), sepiolite (Silva et al., 2021), rice straw biochar (Liu et al., 2018), and shell (Li and Zhang, 2020) can immobilize microorganism for significantly improving the performance of microbe-mineral interaction Bio-PRBs. Various types of Bio-PRBs have been successfully developed for the remediation of groundwater and soil contaminated with atenolol, giferozil, and ciprofloxacin (Vijayanandan et al., 2018), 1,1,1-trichloroethane (1,1,1-TCA) (Wang and Wu, 2017), 1,4-dioxane (Yang et al., 2018), tetracycline (TC) (Huang et al., 2017), Cr(VI) (Huang et al., 2021), TCE (Siggins et al., 2021), Cu(II) and Ni(II) (Silva et al., 2021), nitrobenzene (NB) (Liu et al., 2018), 2,4,6-trichlorophenol (2,4,6-TCP) (Wang et al., 2022), V(V) (Li and Zhang, 2020) (Table 1.1). Furthermore, abiotic reactive materials including zeolite (Vignola et al., 2011), clay minerals (De Pourcq et al., 2015), bimetal (Eljamal et al., 2020), CaO_2 (Liu et al., 2018), MgO_2 (Gholami et al., 2019), etc., have been shown to promote the removal of NB (Liu et al., 2018), cesium-137 (Cs-137) (De Pourcq et al., 2015), phosphorus (Eljamal et al., 2020), and naphthalene and toluene (Gholami et al., 2019) in Bio-PRBs.

According to the properties of contaminants, the microbe-mineral interaction Bio-PRBs can be designed to be single or multiple microorganism-augmented barriers for meeting the requirements (Liu et al., 2016; Upadhyay et al., 2018). In addition, these reported microbial populations (e.g. *Mycobacteria*, *pseudomonas*, and *sphingomonas* (Liu et al., 2019), sulfate-reducing bacteria (*Desulfitobacterium hafniense Y-51* and *Geobacter metallireducens GS-15*) (Wilopo et al., 2008b), *Desulfitobacterium*, *Sulfurospirillium* and *Desulfuromonas* (Siggins et al., 2021), nitrobenzene (NB)-degrading bacteria (Liu et al., 2018) and arsenate-reducing bacterium (*Clostridium*) (Wilopo et al., 2008a)) have been added in microbe-mineral interaction Bio-PRBs for efficiently removing specific contaminants (e.g. PHE, TCE, NB, As(V), and As(III)).

Table 1.1 The reaction parameters and remediation performance of microorganism-activated Bio-PRBs and microorganism-immobilized Bio-PRBs for ordinary and emerging contaminants.

Contaminant type	Influent concentration ($mg \cdot L^{-1}$)	Microbial type	Microorganism activation material	Microorganism immobilized carrier	Other material	Running time (d)	Reaction rate k (d^{-1})	Removal efficiency η (%)	Reference
Cr(VI)	100	Anaerobic bacteria	Pine bark compost	—	Granite powder	1.25	0.159 – 0.493	18 – 46	(Cancelo-González et al., 2015)
Mn(II)	60	*Desulfitobacterium hafniense Y-51*, *Geobacter*, *metallireducens*, *Clostridium sp.* and *OhILAs*	Sheep manure and compost	—	Woodchips, glass beads and ZVI	284	0.003	55	(Wilopo et al., 2008b)
1,1,2-trichloroethane (1,1,2-TCA)	170	*Desulfitobacterium* and *Dehalococcoides*	Degradable organic carbon	—	ZVI	888	0.003 – 0.004	96 – 98	(Patterson et al., 2016)
Benzene	50	Aerobic bacteria	CaO_2 nanoparticles immobilized with sodium alginate	—	—	40	0.230	100	(Mosmeri et al., 2017)
Phenanthrene (PHE)	0.9	*Mycobacteria*, *pseudomonas* and *sphingomonas*	Coconut shell biochar	—	Diatomite, attapulgite and CaO_2	450	1.380	97	(Liu et al., 2019)
o-nitrochlorobenzene (*o*-NCB)	10	*Firmicutes*, *Bacteroidetes* and *Proteobateria*	Wheat straw	—	Attapulgite, diatomite and ZVI	460	0.200	100	(Liu et al., 2021)

(continued)

Contaminant type	Influent concentration ($mg \cdot L^{-1}$)	Microbial type	Microorganism activation material	Microorganism immobilized carrier	Other material	Running time (d)	Reaction rate k (d^{-1})	Removal efficiency η (%)	Reference
As(III)	50	*Desulfitobacterium*, *hafniense Y-51*, *Geobacter* and *metallireducens*	Sheep manure and compost	—	Woodchips, glass beads and ZVI	284	0.006	80	(Wilopo et al., 2008b)
As(V)	5	Sulfate-reducing bacteria (SRB)	Glycerin	—	ZVI and sediment	200	0.039	99.9	(Kumar et al., 2016)
As(V)	6.2	*Desulfuromonas*	Compost leaf mulch, wood chips and organic-rich sediments	—	Limestone and ZVI	210	0.033	99.9	(Angai et al., 2022)
Se(VI)	40	SRB and selenate-reducing bacteria	Municipal leaf compost, sawdust and wood chips	—	Silica sand, gravel, stream sediment and ZVI	60	0.428	99.9	(Sasaki et al., 2008)
Tetracycline (TC)	18	*Anaerolinaceae* and *Hyphomicrobiaceae*	Anaerobic activated sludge	—	ZVI	110	0.021	50	(Huang et al., 2017)
Cr(VI)	50	*S. saromensis W5*	—	Activated carbon	—	120	0.009	60	(Huang et al., 2021)
Cr(VI)	50	*S. saromensis W5*	—	ZVI	—	120	0.019	95	(Huang et al., 2021)
Cr(VI)	50	*S. saromensis W5*	—	Quartz	—	120	0.003	30 – 50	(Huang et al., 2021)
Cr(VI)	132.7	*S. equisimilis*	—	Sepiolite	—	4	0.030	83.3	(Silva et al., 2021)

(continued)

Contaminant type	Influent concentration ($mg \cdot L^{-1}$)	Microbial type	Microorganism activation material	Microorganism immobilized carrier	Other material	Running time (d)	Reaction rate k (d^{-1})	Removal efficiency η (%)	Reference
V(V)	10	*Bacteroides and Geobacter*	—	Shell	Woodchips and sulfur (S (0))	135	1.227	68.5 – 98.2	(Li and Zhang, 2020)
V(V)	10	*Bacteroidetes-vadin HA17 and Geobacter*	—	Quartz sand	S(0)	276	0.075	85	(Shi et al., 2020)
V(V)	10	*Gammaproteobacteria, Thiobacillus, hetrotrophic, Bacteroides and Azotobacter*	—	Woodchips	S(0) and sludge	140	0.106	99.5	(Liu et al., 2022)
U(Ⅵ)	10	*Thiobacillus, hetrotrophic, Bacteroides and Azotobacter*	—	Woodchips	S(0) and sludge	140	0.033	80.7	(Liu et al., 2022)
U(Ⅵ)	10	*SRB and anaerobic bacterial communities with U (Ⅵ)-removal bacteria*	—	Sewage sludge, sawdust and ZVI	Activator aqueous solution, gravel and bone meal	90	0.102	100	(Kornilovych et al., 2018a)
Se(Ⅵ)	0.1	*E.coli, SRB and Se-reducing bacteria*	—	Sediment	Mulch, manure, gravel, limestone and bone meal	70	0.056	98	(Luek et al., 2014)
1,1,1-trichloroethane (1,1,1-TCA)	14.4	*Syntrophobacteraceae*	—	High-density natural luffa sponge	ZVI and sand	288	0.017	99.2	(Wang et al., 2019)

(continued)

Contaminant type	Influent concentration ($mg \cdot L^{-1}$)	Microbial type	Microorganism activation material	Microorganism immobilized carrier	Other material	Running time (d)	Reaction rate k (d^{-1})	Removal efficiency η (%)	Reference
1,1,1-trichloroethane (TCA)	70	*Dechlorinating bacteria(DhB), SRB and Iron-reducing bacteria (IRB)*	—	ZVI	—	280	0.014	97.8	(Yang et al., 2018)
Trichloroethylene (TCE)	1	*Anaerobic TCE degrading bacteria*	—	Commercial compost	—	181	1.200	100	(Ozturk et al., 2012)
Trichloroethylene (TCE)	35	*Desulfitobacterium, Sulfurospirillium and Desulfuromonas*	—	Herbal fruit residue biochar	—	—	0.580	99.7	(Siggins et al., 2021)
Trichloroethylene (TCE)	35	*Desulfitobacterium, Sulfurospirillium and Desulfuromonas*	—	Spruce biochar	—	—	0.580	99.7	(Siggins et al., 2021)
Phenanthrene (PHE)	17.8	*Trichoderma longibrachiatum*	—	Nylon sponge	—	80	0.115	100	(Cobas et al., 2013)
Nitrobenzene (NB)	100	*Pseudomonas putida*	—	Rice straw biochar	Polyvinyl alcohol and 10% CaO_2	37	0.129	98.7	(Liu et al., 2018)
2,4,6-trichlorophenol (2,4,6-TCP)	6	*Dechlorination microorganisms*	—	Peanut shell biochar	Calcium alginate (CA) coated ZVI	3	0.264	79	(Wang et al., 2022)
Atenolol Ciprofloxacin Giferozil	1 1 1	*Anaerobic bacteria*	—	Clay composite adsorbent	Glucose	6.25	0.340 0.390 0.220	80 90 75	(Vijayanandan et al., 2018)
1,4-dioxane	70	*DhB* *SRB* *IRB*	—	ZVI	—	280	0.009	92.5	(Yang et al., 2018)

Chapter 2
Microbe-mineral Interaction Technology

2.1 Why do minerals need coupled microorganism augmentation?

The conventional permanent or semi-permanent reactive barrier is mainly constructed by ZVI (Azubuike et al., 2016). ZVI-PRBs system can effectively remove As(III) (Eljamal et al., 2011), U(VI) (Kornilovych et al., 2018b), As(V) and Se(VI) (Sun et al., 2017a), and Ni(II) (Bilardi et al., 2016) etc. and limit the transport of these contaminants (Galdames et al., 2020). This is economical, without additional energy and operation-simple technology (Cao et al., 2020). However, conventional ZVI-PRBs faced two major challenges in long-term application: unknown iron corrosion kinetics and permeability loss (Cao et al., 2020; Guan et al., 2015). Moreover, it may take decades to eliminate the large contaminant plumes.

Currently, microorganism augmentation with conventional PRBs has successfully solved the above problems. For permeability loss issues, microorganisms addition can degrade organic contaminants into small molecules, CO_2 and H_2O, thereby reducing fillers clogging and maintaining the permeability (Huang et al., 2017). For unknown iron corrosion kinetics issues, microbe-mineral interaction Bio-PRBs improved the utilization of fillers and maintained the reactivity through the synergistic promotion of microorganisms and reactive media, in which microorganisms liquefy metal minerals to slow down corrosion and accelerate the e^{-1} transfer (Huang et al., 2022; Kornilovych et al., 2018b). Furthermore, the microorganisms with other reactive media can improve the reaction rate k, removal efficiency η, and complete degradation of contaminant (e.g. 2, 4-dichlorophenol and tetrabromobisphenol A) (Wang et. al., 2022; Wang et al., 2020) and extend the service life (Siggins et al., 2021) (Table 1.1). Microbe-mineral interaction Bio-PRBs are much more cost-efficient than ZVI-PRBs, pumps and treatments technology, due to their higher overall performance and lower frequency of filler replacement and disposal (Bertolini et al., 2021).

In addition, Li and Liu mainly concentrated on the remediation of heavy metals by PRBs (Li and Liu, 2022). Zhang et al. mainly focused on the application of single zeolite in PRBs and discussed the adsorption and ion exchange mechanism between zeolite and contaminants (Zhang et al., 2022). Amoako-Nimako et al. concentrated on nitrate removal by PRBs constructed with organic substrate or ZVI, and the nitrate removal mechanism in a single medium PRBs was discussed (Amoako-Nimako et al., 2021). Andrade et al. mainly discussed the single action mechanism of four fillers (ZVI, activated carbon, zeolite and microorganism), and focused on the organic compounds in soil (Andrade and dos Santos, 2020). Upadhyay et al. introduced the removal mechanism of permeable reactive bio-barriers (PRBB) for nitrate, heavy metals, chlorinated solvents, and hydrocarbons; whereas, arsenic, vanadium, uranium, selenium, and emerging contaminants were not discussed. Moreover, the microbe-fillers interaction was not discussed (Upadhyay et al., 2018).

2.2 Current status of microbe-mineral interaction technology

Currently, microorganism-activated or microorganism-immobilized Bio-PRBs were developed to remove contaminants, such as heavy metals (e.g. Cr(VI) (Silva et al., 2021), Cu(II) and Zn(II) (Ferronato et al., 2016)), monoaromatic hydrocarbons (e.g. *o*-NCB (Liu et al., 2021), PHE (Liu et al., 2019), NB (Liu et al., 2018)), chlorinated hydrocarbons (e.g. TCE (Siggins et al., 2021), 1,1,2-TCA (Ozturk et al., 2012), 1,1,1-TCA (Wang et al., 2019)), metalloid (e.g. As(V) (Angai et al., 2022), Se(VI) (Sasaki et al., 2008), V(V) (Li and Zhang, 2020)), radioactive contaminant (e.g. U(VI) (Kornilovych et al., 2018b)) emerging contaminants (e.g. tetracycline (TC) (Huang et al., 2017), 1,4-dioxane (Yang et al., 2018), and atenolol, ciprofloxacin, and giferozil

(Vijayanandan et al., 2018)). The extensive results showed that the k and remediation efficiency η increased by 40%–2 650% and 25%–400% in both Bio-PRBs mentioned above, respectively (Table 1.1).

Among them, the k of microorganism-activated Bio-PRBs for treating monoaromatic hydrocarbons (e.g. 2,4,6-TCP (Wang et al., 2022), PHE (Liu et al., 2019)), and metalloid (e.g. As(Ⅴ) (Angai et al., 2022) and Se(Ⅵ) (Sasaki et al., 2008)) increased by 175%–2 650% as compared to ZVI-PRBs (Table 1.1). The coconut shell biochar pellets (coconut shell biochar combined with diatomite, attapulgite, and CaO_2) were adopted to construct Bio-PRBs for treating PHE-contaminated groundwater (Liu et al., 2019). Compared to wheat straw pellets Bio-PRBs (0.013 $mg \cdot g^{-1}$), the PHE adsorption capacities (0.035 $mg \cdot g^{-1}$) increased by 169% in coconut shell biochar Bio-PRBs (Liu et al., 2019). The biochar immobilized microorganism and ZVI gel beads were developed to construct Bio-PRBs for treating 2, 4, 6-TCP contaminated groundwater (Wang et al., 2022). Compared to ZVI-PRBs (0.096 d^{-1}), the k (0.264 d^{-1}) of Bio-PRBs increased by 175% (Wang et al., 2022). Likewise, the organic carbon, silica sand, and organic-rich sediments with ZVI were adopted to construct Bio-PRBs for remediating groundwater contaminated with As(V) (Angai et al., 2022). Compared to sediments PRBs (0.0012 d^{-1}), the k in Bio-PRBs (0.033 d^{-1}) increased by 2 650% (Angai et al., 2022). In addition, municipal leaf compost, sawdust, wood chips, silica sand, gravel, and stream sediment with ZVI were adopted to construct Bio-PRBs for remediation of Se(VI) contaminated groundwater (Sasaki et al., 2008). The reaction rate of Se(VI) increased from 0.137 d^{-1} in ZVI-PRBs (Shrimpton et al., 2018) to 0.428 d^{-1} in Bio-PRBs (Sasaki et al., 2008), i.e., increased by 212%.

Compared to microorganism-activated Bio-PRBs, microorganism-immobilized Bio-PRBs were more efficient for treating heavy metals (e.g. Cr(VI) (Silva et al., 2021)), chlorinated hydrocarbons (e.g. TCE (Siggins et al., 2021)), metalloid (e.g. V(Ⅴ) (Li and Zhang, 2020)), and emerging contaminants (e.g. TC (Huang et al., 2017)) (Table 1.1). Silva et al. (2021) adopted sepiolite-immobilized *S. equisimilis* to construct Bio-PRBs. Compared to sepiolite PRBs (0.0087 d^{-1}), the k in sepiolite immobilized *S. equisimilis* Bio-PRBs (0.03 d^{-1})

increased by 245% (Silva et al., 2021). Siggins et al. adopted biofilm with spruce biochar to construct Bio-PRBs for treating groundwater contaminated with TCE. The k in biofilm with spruce biochar Bio-PRBs (0.58 d^{-1}) of TCE increased by 164% compared to that in spruce biochar PRBs (0.22 d^{-1}) (Siggins et al., 2021). Moreover, Li and Zhang adopted woodchips, sulfur (S(0)), and shell inoculated with the anaerobic microorganism with quartz sand to construct mixotrophic Bio-PRBs for remediating groundwater contaminated with V(V) (Li and Zhang, 2020). Compared to ZVI-PRBs (0.095 d^{-1}) (Morrison et al., 2002), the k (1.227 d^{-1}) of V(V) increased by 1192% in mixotrophic Bio-PRBs (Li and Zhang, 2020). Furthermore, anaerobic-activated sludge and ZVI were adopted to construct Bio-PRBs for remediating groundwater contaminated with TC (Huang et al., 2017). Compared to ZVI-PRBs (40% and 0.015 d^{-1}) and alone microorganism PRBs (10% and 0.0032 d^{-1}), the η (50%) and k (0.021 d^{-1}) of TC increased by 25%-400%, 40%-556% in Bio-PRBs, respectively (Huang et al., 2017).

2.3 Microbe-mineral interaction technology for eliminating different types of contaminants

Both microorganism-activated and microorganism-immobilized Bio-PRBs are appropriate alternative microbe-mineral interaction technologies. Microorganisms and filling materials synergistically accelerate the removal of contaminants and improve the k of microbe-mineral interaction Bio-PRBs.

Currently, many novel microorganisms and filling materials have been tested for optimizing Bio-PRBs with considerable effects on different types of contaminants and applications. According to the collection and classification results of relevant study literature, it was found that microbe-mineral interaction Bio-PRBs technology was mainly focused on eliminating heavy metals, chlorinated hydrocarbon, emerging contaminants, etc.

In this book, the original insights into the different types of active media and microbial carriers, the design and reaction rate of Bio-PRBs, and the interaction mechanisms among 30 species of microorganisms, active media, and comprehensive and representative contaminants were provided (Table 1.1). The typical contaminants were selected and the detailed comparison and discussion are provided as follows chapters.

Chapter 3
Microbe-mineral Interaction for Eliminating Heavy Metals, Metalloid, and Radioactive Substance Contamination

3.1 Heavy metals

Heavy metals are common contaminants of soil and groundwater, which are mainly released into the environment from factories and waste disposal sites, etc. (Pawluk et al., 2015). For heavy metals contaminated groundwater and soil, ZVI-PRBs are the common remediation technology (Jia et al., 2021; Mayacela-Rojas et al., 2021). Whereas, ZVI can cause corrosion products to precipitate and accumulate at the bottom of PRBs, resulting in the reduction of porosity and permeability (Mayacela Rojas et al., 2017). To resolve the problems, novel Bio-PRBs systems were developed.

The vermiculite clay and sediment (vermiculite PRBs), or vermiculite immobilized *pseudomonas putida* and fresh sediment (*pseudomonas* Bio-PRBs) were installed to construct Bio-PRBs columns, respectively, for remediating simulated Cu(II) and Zn(II) contaminated groundwater (Ferronato et al., 2016). Compared to the vermiculite PRBs, the η *of* Cu(II) and Zn(II) increased by 34.4% and 22.8% in *pseudomonas* Bio-PRBs, respectively. The electrostatic attraction of Cu(II) was lower than that of Zn(II) and was not easy to adsorb. Additionally, *pseudomonas putida* acted as a catalyst by secreting exopolysaccharides molecules (EPS) in the adsorption process for promoting the electrostatic attraction between vermiculite clay and Cu(II), Zn(II) to obtain greater η (Ferronato et al., 2016).

Moreover, Huang et al. adopted activated carbon (AC-PRBs), or ZVI (ZVI-

PRBs), or AC immobilized *S. saromensis* W5 (AC Bio-PRBs), or ZVI immobilized *S. saromensis* W5 (ZVI Bio-PRBs) to install four Bio-PRBs columns for treating simulated groundwater contaminated with Cr(VI) (Huang et al., 2021). Compared to AC-PRBs (10.7%), the η of Cr(VI) (64%) increased by 498% in AC Bio-PRBs. The η of ZVI Bio-PRBs (90%) increased by 466% for Cr(VI) compared to that of ZVI-PRBs (15.9%). In ZVI Bio-PRBs, microorganisms, and ZVI synergistic effect significantly decreased the redox potential, providing a reductive environment for microorganisms. The reductive iron bacteria efficiently reduced Fe(III) to Fe(II), providing more electrons to enhance the Cr(VI) reduction (Huang et al., 2021).

Furthermore, the sepiolite, or sepiolite immobilized *S. equisimilis* were adopted to construct two types of Bio-PRBs columns for remediating Cr(VI) contaminated groundwater in chromium plating factory (Figure 3.1) (Silva et al., 2021). Compared to the sepiolite PRBs ($11.7\ mg \cdot g^{-1}$), the adsorption ability ($26\ mg \cdot g^{-1}$) of Cr(VI) in sepiolite immobilized *S. equisimilis* Bio-PRBs increased by 122%; and the removal rate ($0.03\ d^{-1}$) of sepiolite immobilized *S. equisimilis* Bio-PRBs increased by 233% than that of sepiolite PRBs ($0.009\ d^{-1}$). Cr(VI) was removed through surface complexes (H_2O and hydroxyl radicals ($\cdot OH$)) on the surface of sepiolite, and produced Cr^{6+} ions were transformed by ion exchange with Mg^{2+} ions in the lattice of sepiolite (Figure 3.1). Furthermore, Cr(VI) was also removed by surface functional group adsorption of *S. equisimilis* in Bio-PRBs (Figure 3.1) (Silva et al., 2021). Sepiolite immobilized *S. equisimilis* were adopted to construct the pilot-scale Bio-PRBs cylinder for treating heavy metals and organic solvents contaminated groundwater in the chromium plating factory site. The η of total Cr reached 83.3%; diethylketone and methyl ethyl ketone were removed at 96.0% and 86.0%, respectively.

In summary, coupling the microorganisms and electron donor group of organic matter or ZVI accelerated the reduction process of heavy metals, and heavy metals were adsorbed through the functional groups on biomass. However, compared with one heavy metal or multiple heavy metals, co-contamination of heavy metals and organic substances (such as polycyclic aromatic hydrocarbons

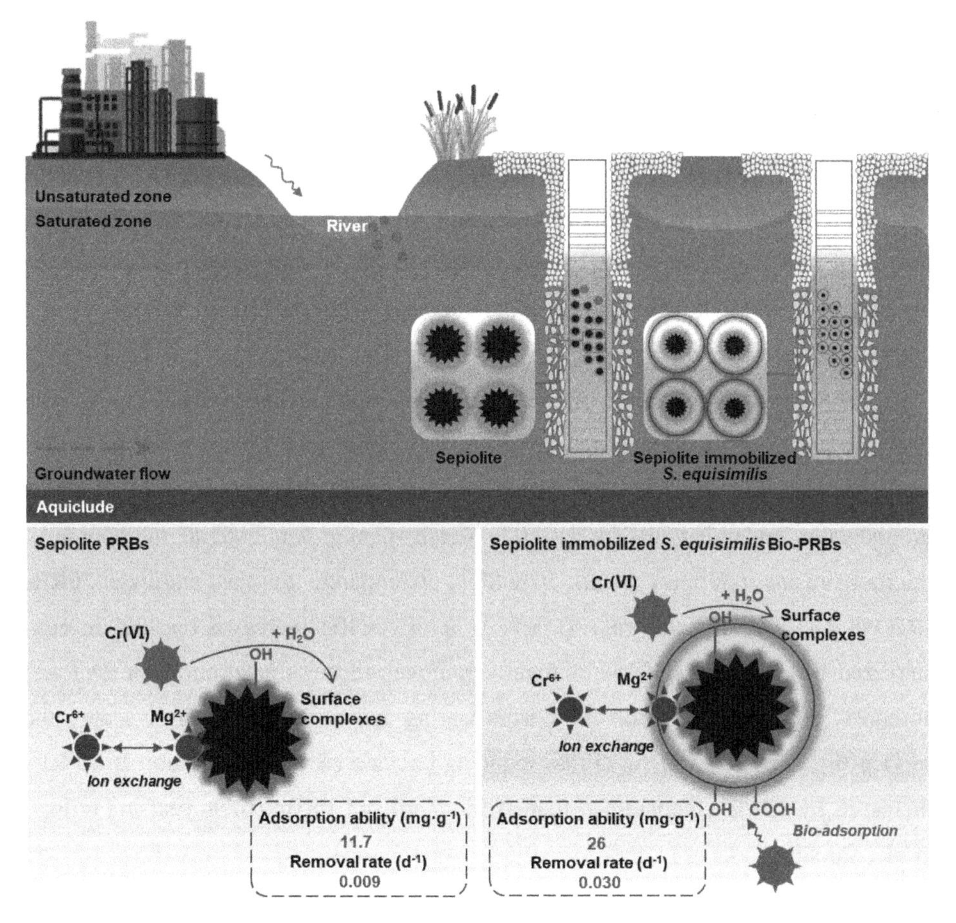

Figure 3.1 Removal performance and reaction mechanism of Cr(VI) in PRBs constructed by sepiolite; Bio-PRBs constructed by sepiolite immobilized *S. equisimilis* biofilm (Silva et al., 2021).

(PAHs)) needs more research. In the future, the filler materials shall be modified for specific contaminants based on the actual composite contamination situation.

3.2 Arsenic

At present, the arsenic (As) content in groundwater is high in mining and surrounding areas, and it poses a threat to human health due to its toxicity (Li et al., 2021; Ouyang et al., 2020). Human activities such as pesticide production, and industrial waste can release As into ecosystems, and most soluble As substances can enter surface water through runoff and leaching into groundwater (Altowayti et al., 2021). The WHO set that the As concentration in drinking water should not exceed 10 $\mu g \cdot L^{-1}$ due to its high toxicity (Öztürk et al., 2017). As in water is mainly present as inorganic As(III) (AsO_3^{3-}) or As(V) (AsO_4^{3-}), which can cause keratosis, hyperpigmentation, and cancer (Kumari and Maurya, 2019; Ren et al., 2017). The results showed that Bio-PRBs significantly improved the remediation effect of groundwater contaminated with As(III) (Wilopo et al., 2008b) or As(V) (Angai et al., 2022; Kumar et al., 2016).

The non-sterilized sheep manure, compost, woodchips, glass beads with ZVI, or sterilized sheep manure, compost, woodchips, and glass beads with ZVI were adopted to construct two Bio-PRBs columns for remediating simulated groundwater contaminated with As(III). Sheep manure was used as the source of microorganisms (Wilopo et al., 2008b). Compared to the sterilized PRBs (0.0038 d^{-1}), the reaction rate (0.0057 d^{-1}) of As(III) increased by 50% in non-sterilized Bio-PRBs. As(III) was mainly removed through adsorption on ZVI and compost, and coprecipitation with iron-bearing minerals. On the full scale, the service life (15.76 years) of sulfate-reducing bacteria (SRB) inoculation Bio-PRBs increased by 134% compared to that of the sterilized PRBs (6.74 years) (Wilopo et al., 2008b).

Moreover, the glycerin, ZVI with non-sterilized sediment mixture, or glycerin, ZVI with sterilized sediment mixture were used to construct two Bio-PRBs columns for treating simulated groundwater contaminated with As(V). Glycerin acts as a stimulant of local microorganisms to enhance activity (Kumar et al., 2016). Compared to the sterilized sediment PRBs (5.6%), the removal rate (50%) of total As increased by 793% in non-sterilized sediment Bio-PRBs. After 90 d treatment, the concentration of remaining As(V) was below the WHO acceptable limit (10 $\mu g \cdot L^{-1}$) in Bio-PRBs. As(Ⅲ) was the main form of As in non-sterilized sediment Bio-PRBs, however, almost all As was As(V) in sterilized sediment PRBs. As(V) obtained electrons to form As(Ⅲ) and precipitated as AsS. Additionally, As(V) was also removed through the adsorption of iron (hydrogen) oxides (Kumar et al., 2016).

Furthermore, Angai et al. adopted silica sand with organic-rich sediments (organic-rich sediments Bio-PRBs), or silica sand, organic-rich sediments, and organic carbon with ZVI (organic-rich sediments + ZVI Bio-PRBs) to construct two Bio-PRBs columns for remediating As(V) contaminated groundwater. Sediments were the source of microorganisms, and organic carbon was used to stimulate the growth and activity of microorganisms (Figure 3.2) (Angai et al., 2022). Compared to organic-rich sediments Bio-PRBs (22.6%), the η (99.9%) of total influent As(V) increased by 342% in organic-rich sediments + ZVI Bio-

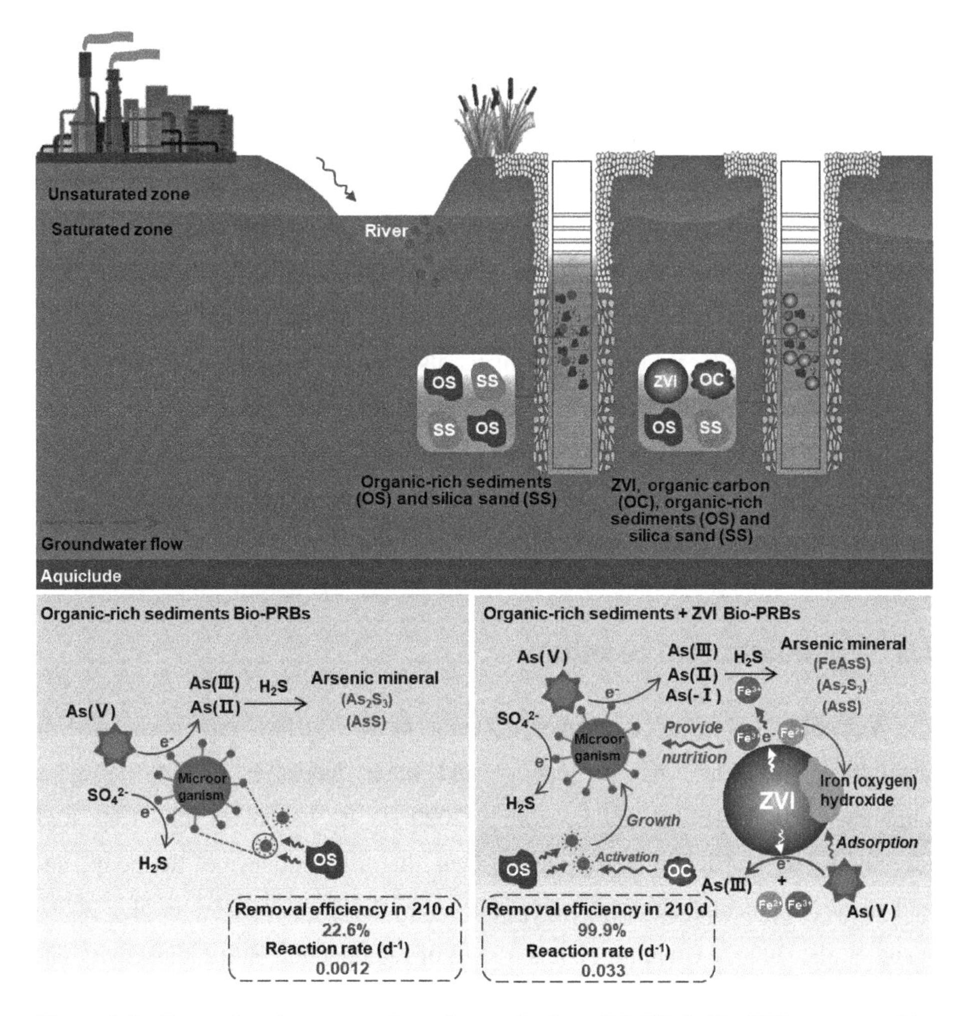

Figure 3.2 Removal performance and reaction mechanism of As(V) in Bio-PRBs constructed by organic-rich sediments with silica sand; and Bio-PRBs constructed by organic carbon (compost leaf mulch, wood chips), silica sand, and organic-rich sediments with ZVI (Angai et al., 2022).

PRBs. The *k* (0.033 d^{-1}) of organic-rich sediments + ZVI Bio-PRBs increased by 2 438% compared to that of organic-rich sediments Bio-PRBs (0.0013 d^{-1}). Sulfate-reducing bacteria (SRB) reduced SO_4^{2-} to H_2S, which transform As(Ⅲ) and As(Ⅱ) to form As_2S_3 and AsS precipitation in organic-rich sediments Bio-PRBs (Figure 3.2). However, in organic-rich sediments + ZVI Bio-PRBs, As(V) obtained electrons to form As(Ⅲ), As(Ⅱ), and As(−Ⅰ) for forming As_2S_3, AsS and FeAsS precipitation. ZVI promoted the reduction of As(V), and iron (oxygen) hydroxide removed part of As(V) through adsorption. Fe(III) as an electron acceptor was utilized by SRB and promoted precipitation formation (Figure 3.2) (Angai et al., 2022).

Generally, compared to conventional PRBs, Bio-PRBs are more effective (*k* increased by 50%−2 438%) for As(V) and As(Ⅲ) removal. Moreover, organic matter from organic-rich sediments enhanced the growth of SRB and maintained its high activity, so that SO_4^{2-} can be continuously reduced to reductant H_2S. Subsequently, As can be removed by adsorption, reduction, and coprecipitation with sulfide. Moreover, iron-based materials (such as pyrite and FeS_2) are gradually coming into the view of researchers. Studies have proved the interaction and "mutual benefit" relationship between iron-based minerals and microorganisms. Therefore, the application of "microbe-mineral" interaction in Bio-PRBs is of great significance and worth exploring.

3.3 Vanadium and uranium

Vanadium (V), as a catalyst (Zhang et al., 2019) has been used in metallurgy, medicine, nuclear energy and other industries due to its good toughness and ductility (Zhu et al., 2020). It becomes toxic at high concentrations, and it can affect plant growth, harm the kidneys of the body, and lead to developmental abnormalities, and even death (Aihemaiti et al., 2020; Wang et al., 2020). Uranium (U) is a kind of radionuclide with radioactivity and toxicity, and it enters the environmental medium mainly through metallurgical and mining operations. Moreover, U can threaten human health through the food chain (Chen et al., 2021; Madejón et al., 2022). The permitted V level in drinking water is 0.2 μg · L^{-1} by the US Environmental Protection Agency (USEPA) (Chen and Liu, 2017). To date, Bio-PRBs have been actively used to remove V(V) (Li and Zhang, 2020; Liu et al., 2022; Shi et al., 2020) and U(Ⅵ) (Kornilovych et al., 2018b) from contaminated groundwater.

Li and Zhang adopted woodchips, sulfur (S(0)), shell immobilized with the anaerobic consortium, and quartz sand to construct a Bio-PRBs (mixotrophic Bio-PRBs) reactive column for remediation of V(V) contaminated groundwater. The S(0) was the electron donor, and woodchips as an organic source maintained the activity of microorganisms. Quartz sand was added to improve the permeability of Bio-PRBs (Figure 3.3) (Li and Zhang, 2020). Compared to ZVI-PRBs (0.095 d^{-1}) (Morrison. et al., 2002), the removal rate k (1.227 d^{-1}) of V(V) increased by 129% in mixotrophic Bio-PRBs (Li and Zhang, 2020). The transformation of V(V) in ZVI-PRBs mainly involved insoluble V(Ⅳ) and V(Ⅲ) produced in the ZVI reduction process; V(V) was also adsorbed by iron (oxygen) hydroxide (Figure 3.3) (Morrison. et al., 2002). By contrast, in mixotrophic Bio-PRBs, the S(0) was oxidated by the autotrophs, and SO_4^{2-} was released; V(V) was reduced simultaneously. Whereas, heterotrophs used woodchips as the organic source to complete V(V) biological reduction and OH^- release. Autotrophs and heterotrophs worked together to neutralize the pH of the medium and maintain an environment suitable for microbial life (Figure 3.3) (Li and Zhang, 2020).

Moreover, Shi et al. adopted an anaerobic microbial consortium and S(0) with quartz sand to construct a Bio-PRBs reactive column for treating simulated V(V) contaminated groundwater. S(0) contributed electrons, quartz sand regulated the permeability of Bio-PRBs (Shi et al., 2020). In Bio-PRBs, the greatest k of V(V) reached 0.075 d^{-1} in 166 – 209 d, increased by 10% – 159% among other period (0.029 d^{-1}, 0.068 d^{-1}, 0.058 d^{-1}, respectively). Additionally, insoluble V(Ⅳ) formed from V(V) reduction, with significantly reduced mobility and toxicity. At the genus level, functional microorganisms (*Bacteroidetes-vadin HA17* and *Geobacter*) of reduction and oxidation accelerated the transformation of contaminants (Shi et al., 2020).

Liu et al. adopted woodchips and S(0) with sludge to construct a Bio-PRBs

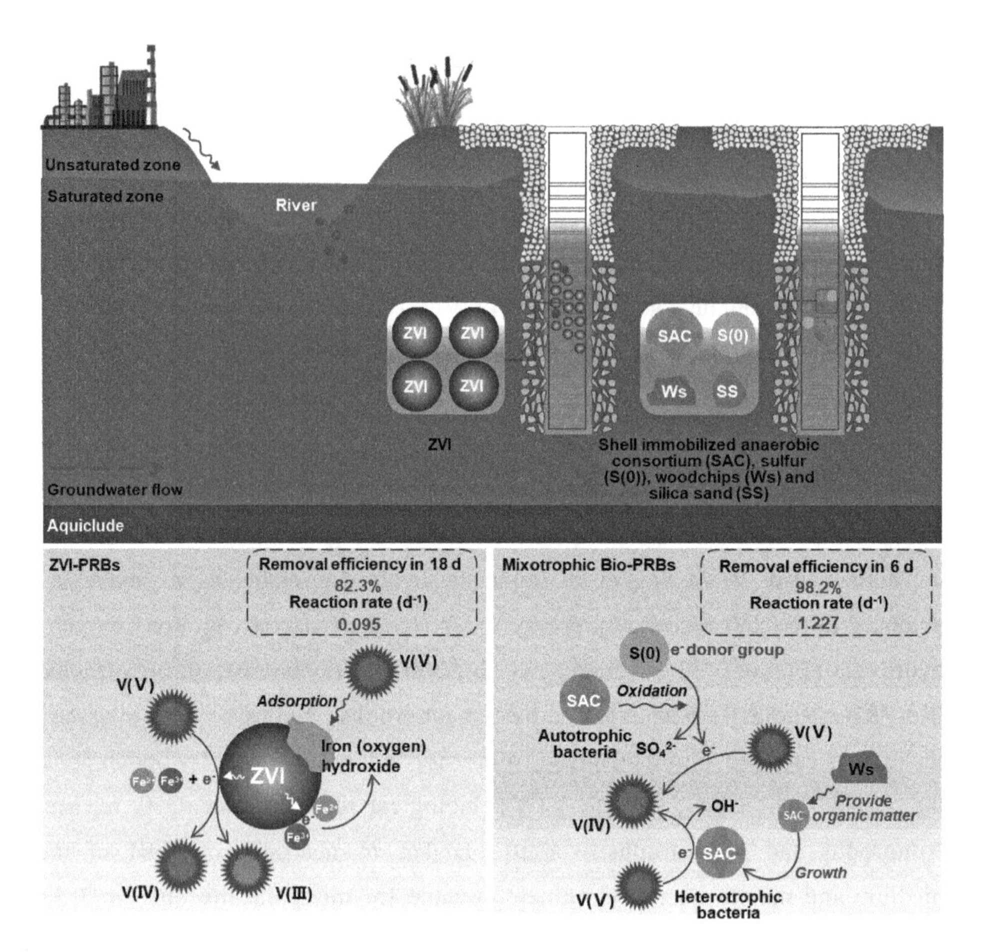

Figure 3.3 Removal performance and reaction mechanism of V(Ⅴ) in PRBs constructed by ZVI (Morrison. et al., 2002); Bio-PRBs constructed by woodchips, S(0), shell immobilized with the anaerobic consortium, and quartz sand (Li and Zhang, 2020).

column for remediation of simulated groundwater contaminated with V(V) and U(Ⅵ). Woodchips were a carbon source and electron donors for specific microorganisms and S(0) acted as electron donors (Liu et al., 2022). In Bio-PRBs, the greatest k of V(V) reached 0.106 d^{-1} in 0–50 d, increased by 166%–290% among other period (0.0272 d^{-1}, 0.0369 d^{-1}, 0.0399 d^{-1}, respectively). Moreover, the η of V(V) and U(VI) simultaneously reached 99.5% and 80.7% in Bio-PRBs, respectively (Liu et al., 2022). V(V) was reduced in preference to U(Ⅵ) due to its high redox potential. Insoluble and less toxic $VO(OH)_2$ and UO_2

formed during the reduction process. Compared with U(Ⅵ), microorganisms can better adapt to V(V) and the abundance of *Gammaproteobacteria* increased from 19.49% to 28.44%, in which *Gammaproteobacteria* promoted the V(V) reduction. Additionally, it's worth noting that some bacteria, such as *Thiobacillus*, *Bacteroides*, and *Azotobacter*, can achieve simultaneous removal of both contaminants (Liu et al., 2022).

Furthermore, Kornilovych et al. adopted microorganisms and other fillers (sewage sludge, bone meal, sawdust, activator aqueous solution, and gravel), or microorganisms, and other fillers with ZVI to construct two Bio-PRBs for treating real groundwater contaminated with U(VI). An activator aqueous solution was added to activate microorganisms for facilitating the process of bio-reduction (Kornilovych et al., 2018b). Results showed that U(VI) was almost completely removed during 90 d, i.e., k was 0.102 d^{-1}. U(VI) was removed by binding with ligands such as ethylenediaminetetraacetic acid (EDTA) and fulvic acid (FA). Moreover, organic active filling-sewage sludge (source of microorganisms), bone meal, sawdust, indigenous bacteria culture and gravel were adopted to construct the Bio-PRB cylinders for the remediation of the actual U(VI) contaminated groundwater in the Shch tailings facility site. The U(VI) concentration (0.38 to 0.07 $mg \cdot L^{-1}$) significantly decreased, corresponding to 81.6% removal efficiency (Kornilovych et al., 2018b).

In conclusion, the coupling of sludge (microorganism) and electron donor group S(0) significantly improved k of Bio-PRBs (increased by 10%-290%). Microorganisms promoted the formation of insoluble and less toxic V(Ⅳ) and U(Ⅳ). For the removal of V(V) and U(VI), the reduction process is a critical step (reduction of toxicity and solubility). In addition, Future research can focus on the radioactivity impact of V(V) and U(VI) on the activity of microorganisms, and explore the relationship between V(V) and U(VI) radioactivity and microbial activity to optimize Bio-PRBs.

3.4 Selenium

Selenium (Se) is an essential nutrient (Hawrylak-Nowak et al., 2015). It was used in ceramics, metallurgy, glass industries, and other industries (Okonji et al., 2021). Se is beneficial in reducing heart disease, maintaining the normal function of the heart, and preventing cancer (Hawrylak-Nowak et al., 2015). However, it becomes toxic in high concentrations and may cause brain diseases and cancer (Kursvietiene et al., 2020). The WHO suggested that the average daily intake of Se is about 30–50 $\mu g \cdot d^{-1}$(Kieliszek and Blazejak, 2016). Inadequate or excess Se intake will cause adverse effects on the human body (safety range 40–400 $\mu g \cdot d^{-1}$) (Chen et al., 2019; Vogel et al., 2018). Some researchers tried to remediate Se(VI) (Sasaki et al., 2008) and Se(Ⅳ) (Luek et al., 2014) contaminated groundwater with Bio-PRBs and reached a satisfying removal effect.

Municipal leaf compost, sawdust, wood chips, silica sand, gravel, and stream sediment with ZVI were adopted to construct a Bio-PRBs column for remediation of Se(VI) contaminated groundwater. Stream sediment was added as the anaerobic bacteria source (Sasaki et al., 2008). Compared to ZVI-PRBs (35%) (Shrimpton et al., 2018), the η (95%) of Se(VI) in ZVI + sediment Bio-PRBs increased by 171%; and the k (0.428 d^{-1}) of ZVI + sediment Bio-PRBs (Sasaki et al., 2008) increased by 212% compared to that of ZVI-PRBs (0.137 d^{-1}) (Figure 3.4) (Shrimpton et al., 2018). Se(VI) was adsorbed on iron (oxygen) hydroxide and mainly reduced to iron selenide (FeSe and $FeSe_2$) and insoluble element Se (hexagonal types) in ZVI-PRBs (Figure 3.4) (Shrimpton et al., 2018). However, Se(VI) was adsorbed on iron (oxygen) hydroxide, and reduced to FeSe, $FeSe_2$, and element Se (hexagonal and amorphous types) in ZVI + sediment Bio-PRBs. In addition, microorganisms promoted the reduction of Se(VI) and the liquidation of iron (oxygen) hydroxide, iron ions (Fe^{2+} and Fe^{3+}) released from ZVI and iron (oxygen) hydroxide provided nutrition for microorganisms, thus forming an excellent cycle of iron selenium mineral precipitation (Figure 3.4) (Sasaki et al., 2008).

The different ratios of mulch and manure (type 1 = 40 : 32, type 2 = 57 : 15), bacterial inoculum, gravel, and limestone with the bone meal were adopted to construct two field-scale Bio-PRBs for treating Se(VI) and Se(Ⅳ) contaminated surface runoff in the lower end of the drainage system of the mine.

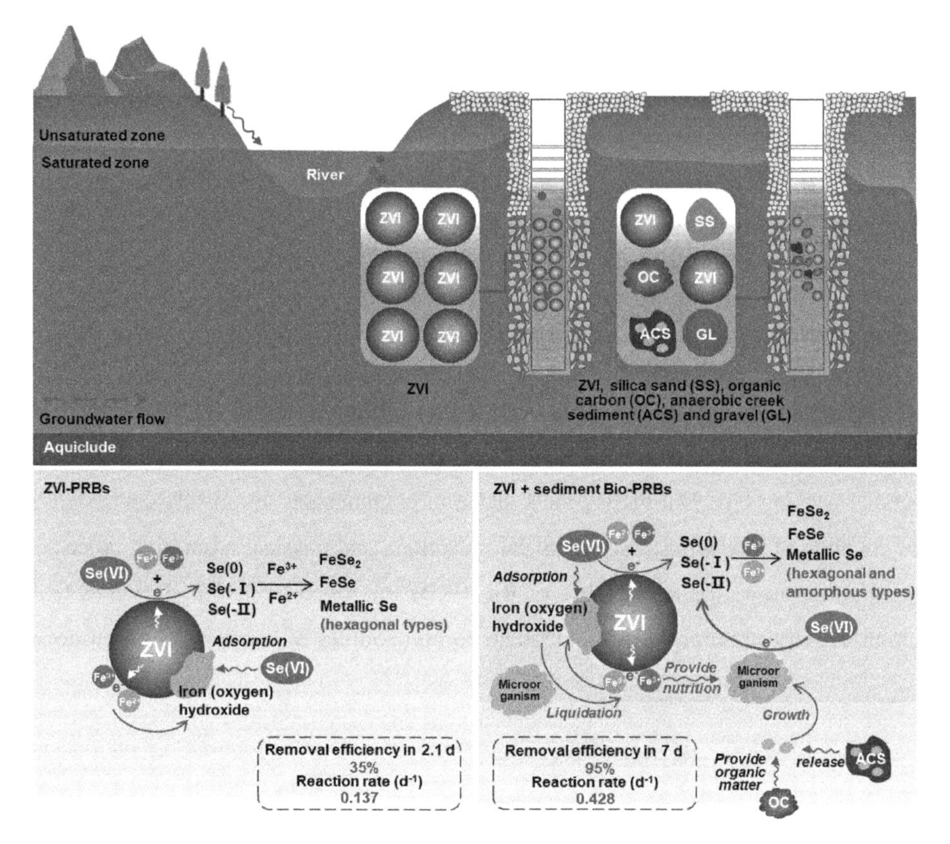

Figure 3.4 Removal performance and reaction mechanism of Se(VI) in PRBs constructed by ZVI (Shrimpton et al., 2018); and Bio-PRBs constructed by ZVI, anaerobic creek sediment, silica sand, and gravel with organic carbon (Sasaki et al., 2008).

Mulch, bone meal, and manure were carbon sources, phosphorus sources and nitrogen sources, respectively (Luek et al., 2014). Compared to type 1 Bio-PRBs (0.04 d^{-1}), k of total Se (Se(VI) and Se(Ⅳ)) increased by 40% in type 2 Bio-PRBs (0.056 d^{-1}). Se(VI) and Se(Ⅳ) acted as an energy source to maintain microorganism activity, which was continuously reduced to the element Se and formed a good cycle. Moreover, manure acted as both a bacterial carbon source and nitrogen source, and *E. coli* from manure enhanced the bio-reduction process. Compared to type 1 Bio-PRBs, in the type 2 Bio-PRBs, 57% of manure released more nitrogen that was necessary for microbial life, maintained microbial activity, and improved η of Se(VI) and Se(Ⅳ) (Luek et al., 2014).

Taken together, Bio-PRBs can efficiently achieve Se(VI) toxicity reduction.

It promotes the reduction process of Se(VI) to Se(Ⅳ) and element Se. Additionally, microorganisms accelerate the formation of selenide deposits. The interaction between microorganisms and Se(VI) and Se(Ⅳ) needs to be further studied to optimize Bio-PRBs for adjusting the Se(VI) and Se(Ⅳ) concentration within the acceptable range for human beings.

3.5 Conclusions and suggestions

A certain amount of work has been done to improve the k, η, and service life of microbe-mineral interaction Bio-PRBs. In various novel microorganisms with metallic packing type Bio-PRBs, the k and η of contaminants are mainly controlled by the coupling of biodegradation, chemical reduction, and physical adsorption processes.

Furthermore, the service life of Bio-PRBs improved by 134%. It comes from a high-activity microorganism, adequate carbon sources of carriers, electron donor

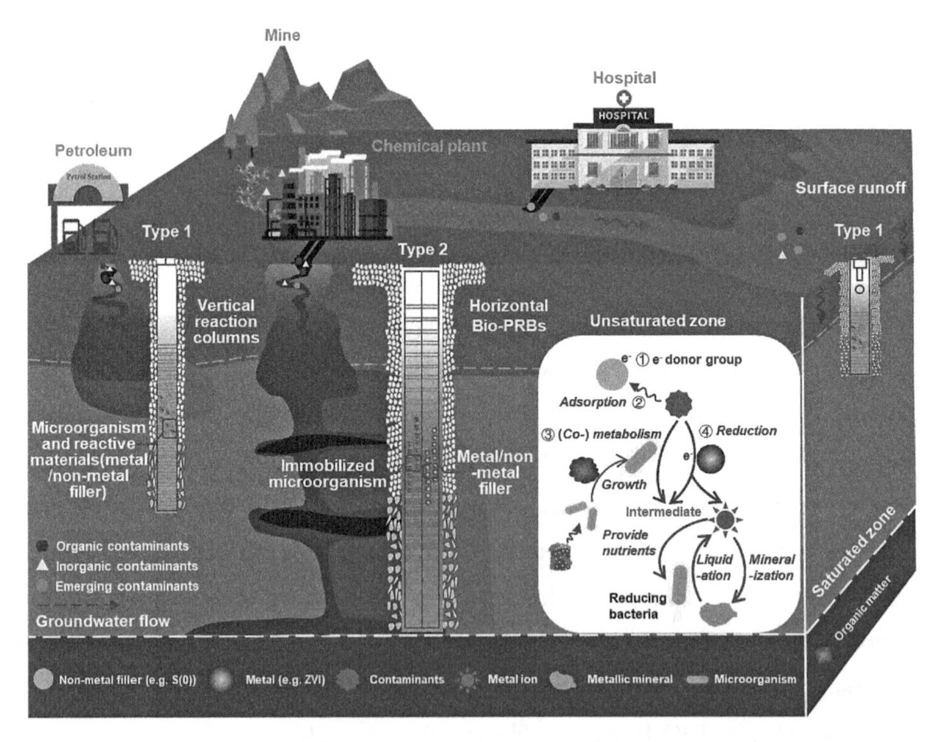

Figure 3.5 Bio-PRBs efficient eliminating contamination of soil and groundwater based on microbe-mineral interaction and stimulation.

groups of ZVI and S(0) etc. Bio-PRBs provided efficient and novel approaches and strategies for eliminating soil and groundwater contamination based on microbe-mineral interaction and stimulation (Figure 3.5).

The above research has demonstrated that microbe-mineral interaction Bio-PRBs are effective, however, some matters need attention. Suggestions on the application and development of Bio-PRBs are proposed. To improve the *k* of Bio-PRBs, novel materials that can stimulate and immobilize microorganisms need to be further explored on a scale; the relationship between the microbial and non-microbial filling material and the service life of Bio-PRBs for treating real groundwater needs to be deeply investigated.

Chapter 4
Microbe-mineral Interaction for Eliminating Organic Contaminants

4.1 Aromatic hydrocarbons

Aromatic hydrocarbons include monoaromatic hydrocarbons, polycyclic aromatic hydrocarbons, and thick cyclic aromatic hydrocarbons. Monoaromatic hydrocarbons are common contaminants produced by human activities, which may threaten human health (Ewlad-Ahmed et al., 2021). Benzene, nitrobenzene (NB) and ethylbenzene, etc. are widely used in the chemical industry to synthesize other products (Sorensen et al., 2019). They are the representative light non-aqueous phase liquids (LNAPL), which are released during the use process and accumulated above the groundwater table (Sookhak Lari et al., 2019). At present, monoaromatic hydrocarbons have been identified by the WHO as a strong carcinogen (Ewlad-Ahmed et al., 2021). The high solubility of monoaromatic hydrocarbons accelerates their transport and contamination risk, which is seriously harmful to human health (Mazzeo et al., 2010).

Currently, researchers have developed various types of Bio-PRBs to treat monoaromatic compounds contaminated soil and groundwater; and the *k* increased by 119%–395% in Bio-PRBs when compared with polyvinyl alcohol and biochar PRBs (Liu et al., 2018) and ZVI-PRBs (Ren et al., 2019). CaO_2 nanoparticles sodium alginate gel beads or free CaO_2 nanoparticles were adopted to construct two Bio-PRBs columns for remediating benzene-contaminated groundwater. CaO_2 nanoparticles acted as a source of O_2 (Mosmeri et al., 2017). Compared to free CaO_2 Bio-PRBs (61%), the biodegradation efficiency (100%) of benzene increased by 64% in encapsulated CaO_2 Bio-PRBs. O_2 released from CaO_2

stimulated aerobic microorganisms' growth for accelerating benzene removal (Mosmeri et al., 2017).

Liu et al. installed polyvinyl alcohol, biochar, and microorganisms (PBM) with 10% CaO_2, or PB (polyvinyl alcohol and biochar) into a solid cube to construct two Bio-PRBs columns for treating simulated NB contaminated groundwater. Biochar acts as a carrier and adsorbent for immobilizing microorganisms and maintaining biological activity (Liu et al., 2018). Compared to PB-PRBs (0.059 d^{-1}), the k (0.129 d^{-1}) of NB increased by 119% in PBM with 10% CaO_2 Bio-PRBs. Encapsulated medium prevented a high concentration of NB from damaging NB-degrading bacteria. After 15–20 d of continuous O_2 release from CaO_2, the dissolved oxygen (DO) concentration still can reach 8.0 mg · L^{-1}. O_2 as the electron acceptor for NB-degrading bacteria promoted the removal of NB (Liu et al., 2018).

Liu et al. adopted the quartz sand with wheat straw biochar gel beads (wheat straw with diatomite, attapulgite, CaO_2) or coconut shell biochar gel beads (coconut shell biochar with diatomite, attapulgite, CaO_2) to construct two Bio-PRBs columns for remediating simulated thick cyclic aromatic hydrocarbon phenanthrene (PHE) contaminated groundwater (Liu et al., 2019). Compared to wheat straw gel beads Bio-PRBs (0.013 mg · g^{-1}), the PHE adsorption (0.035 mg · g^{-1}) increased by 169% in coconut shell gel beads Bio-PRBs (Liu et al., 2019). The k (1.38 d^{-1}) of PHE increased by 395% in coconut shell gel beads Bio-PRBs (Liu et al., 2019) compared to that in ZVI-PRBs (0.279 d^{-1}) (Ren et al., 2019). PHE was mainly removed by the adsorption of iron (oxygen) hydroxide, activated carbon, and reduction of ZVI in ZVI-PRBs (Figure 4.1) (Ren et al., 2019). Whereas, in coconut shell gel beads Bio-PRBs, biodegradation of PHE was significantly enhanced by O_2 from CaO_2 and organic carbon from attapulgite (Figure 4.1). Coconut shell gel beads stably released organic carbon and O_2, which were beneficial to maintain the activity of microorganisms and improve the abundance of microorganisms, thus promoting the biodegradation of PHE. *Mycobacteria*, *pseudomonas*, and *sphingomonas* were the main PHE-degrading bacteria and were conducive to enhancing the release of organic carbon from the gel beads. Moreover, the coconut shell gel beads of Bio-PRBs can be used continuously for at least 12.3 years (Figure 4.1) (Liu et al., 2019).

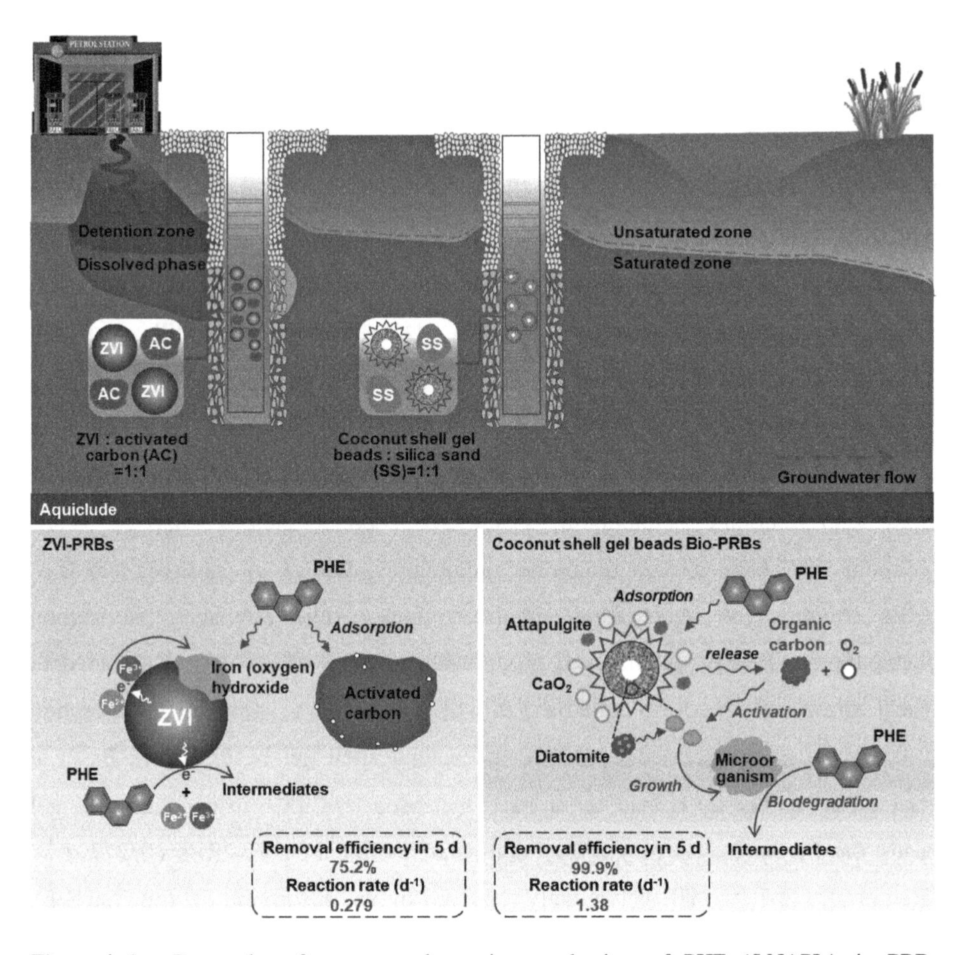

Figure 4. 1 Removal performance and reaction mechanism of PHE (LNAPL) in PRBs constructed by ZVI with activated carbon (Ren et al., 2019); Bio-PRBs constructed by anaerobic microorganism, coconut shell gel beads (coconut shell biochar combined with diatomite, attapulgite, CaO_2) and silica sand (Liu et al., 2019).

Wang et al. adopted ZVI gel beads with biochar immobilized microorganisms gel beads (Bio-PRBs), or only ZVI gel beads (ZVI-PRBs) to construct two Bio-PRBs columns for treating simulated groundwater contaminated with 2,4,6-TCP (Wang et al., 2022). Compared to ZVI-PRBs (0.48), the dispersity *a* (0.2) decreased by 140% in Bio-PRBs, and the k in Bio-PRBs (0.264 d^{-1}) increased by 175% compared to that in ZVI-PRBs (0.096 d^{-1}). 2,4,6-TCP was rapidly degraded through the coupling of the ZVI chemical reduction and biological cometabolism. Furthermore, the field-scale vertical Bio-PRBs reaction columns were constructed to treat actual chlorinated organic compounds and Benzene Toluene Ethylbenzene & Xylene (BTEX) contaminated groundwater in a pesticide factory site. The results demonstrated this type of Bio-PRBs can efficiently remove chlorobenzene, methylbenzene, 1,4-dichlorobenzene, benzene, ethylbenzene, chlorobenzene, 2-chlorotoluene, and the η ranged from 61% – 100% (Wang et al., 2022).

Based on the above points, Bio-PRBs is potential and environmentally friendly for eliminating monoaromatic hydrocarbons. The coupling of the biological cometabolism and ZVI chemical reduction or CaO_2 addition (releasing electron acceptor O_2) is conducive to the degradation of benzene, NB, PHE, 2,4,6-TCP, and other monoaromatic hydrocarbon contaminants into small molecules. Compared with the laboratory, the diversity of microorganisms in the actual groundwater is more abundant, highly efficient transition from the laboratory to the actual site needs to be optimized. The types of oxygen-releasing agents (not only CaO_2,) also need further exploration.

4.2 Chlorinated hydrocarbons

Chlorinated hydrocarbons, such as 1,2,3-TCP, 1,1-dichloroethane (1,1-DCA), 1,1-dichloroethylene (1,1-DCE) and vinyl chloride (VC), are widely applied in chemistry, dyeing and other industries (Samin and Janssen, 2012; Yang et al., 2018). In addition, organic chlorinated hydrocarbon (e.g. TCE) are the most common contaminants of dense non-aqueous phase fluids (DNAPLs) (Engelmann et al., 2021). These contaminants may threaten human health due to their bioaccumulation and toxicity (Arjoon et al., 2012). Currently, many studies have shown that Bio-PRBs can effectively remove TCE (Siggins et al., 2021), 1,1,2-TCA, and VC (Patterson et al., 2016) from groundwater.

Anaerobic TCE degrading microorganisms with organic natural medium eucalyptus mulch, or anaerobic TCE degrading microorganisms with commercial compost were adopted to construct two Bio-PRBs columns for treating simulated TCE-contaminated groundwater. The eucalyptus mulch and commercial compost were used as carbon sources and nutrients for enhancing the growth of microorganisms (Ozturk et al., 2012). Compared to eucalyptus mulch Bio-PRBs ($0.23\ d^{-1}$), the k ($1.2\ d^{-1}$) of TCE increased by 422% in compost Bio-PRBs. Furthermore, the retardation factor R_d (301) of compost Bio-PRBs for TCE increased by 760% compared to that in eucalyptus mulch Bio-PRBs (35). TCE was mainly biodegraded to *cis*-1, 2-DCE, VC, and a small amount of ethylene (Ozturk et al., 2012). Degradable organic carbon with ZVI was adopted to construct a Bio-PRBs column for treating groundwater contaminated with 1,1,2-TCA (Patterson et al., 2016). Organic carbon was used as a carbon source to enhance microorganism activity, and ZVI served as the electron donor. Compared to the influent flow rate of $300\ mL \cdot d^{-1}$ condition (33%), the η of 1,1,2-TCA (98%) increased by 197% in Bio-PRBs under $600\ mL \cdot d^{-1}$ condition. Furthermore, the η of by-product VC of 1,1,2-TCA also reached 73%–81% in the Bio-PRBs. ZVI reduction and *Desulfitobacterium* and *Dehalococcoides* biodegradation were mainly mechanisms for enhancing 1, 1, 2-TCA and VC removal (Patterson et al., 2016). Furthermore, the herbal pomace biochar (HPB) immobilized biofilm, or spruce biochar (SB) immobilized biofilm, or HPB, or SB was adopted to construct four Bio-PRBs columns for remediating simulated TCE contaminated groundwater. The biochar was used as a growth substrate to enhance the dechlorination biofilm formation of TCE (Figure 4.2) (Siggins et al., 2021). Compared to SB PRBs ($0.22\ d^{-1}$), the k ($0.58\ d^{-1}$) of SB Bio-PRBs increased by 164%. In addition, except for biochar adsorption, TCE was also adsorbed and biodegraded through microorganisms, and the degradation products were mainly *cis*-1, 2-DCE in Bio-PRBs (Figure 4.2). Biochar provides a carbon source for microorganisms to maintain their activity, forming a good interaction between microorganisms and biochar for treating TCE-contaminated groundwater (Figure 4.2) (Siggins et al., 2021).

In general, Bio-PRBs are friendly and effective for eliminating chlorinated

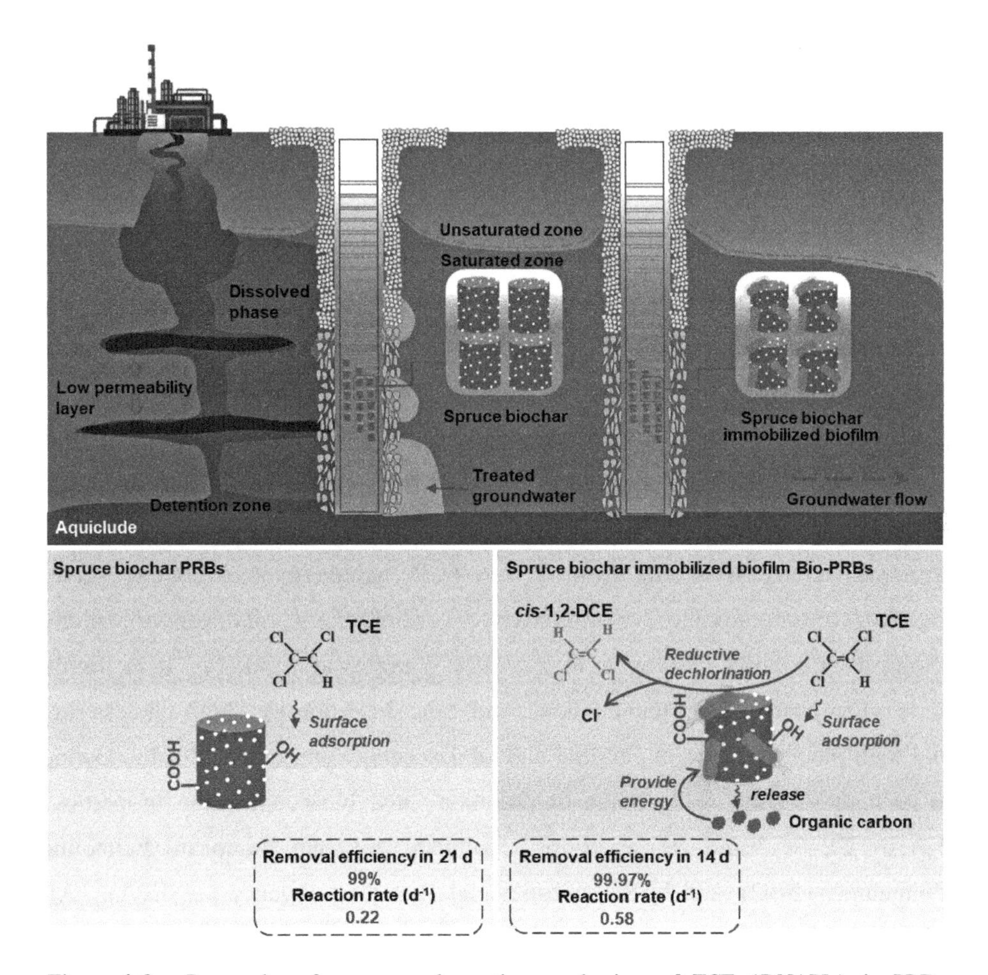

Figure 4.2 Removal performance and reaction mechanism of TCE (DNAPL) in PRBs constructed by spruce biochar; Bio-PRBs constructed by dechlorination microorganisms with spruce biochar (Siggins et al., 2021).

hydrocarbon. The coupling of microorganism dechlorination and ZVI reduction or adsorption through biochar surface functional groups (e.g. -OH and -COOH) significantly improved the k of Bio-PRBs (increased by 164%−422%). However, the types of microorganisms for completely degrading chlorinated hydrocarbons to non-toxic substances still need to be explored. In addition, the service life of Bio-PRBs composed of organic natural medium eucalyptus mulch or biochar is to be determined, which is very significant for practical application.

4.3 Case study for 2,4,6-TCP elimination by microbe-Fe^0 interaction

4.3.1 Research background

According to statistics, more than 50% of urban groundwater has seriously deteriorated in China (Li et al., 2021). In the urban and rural areas, chlorophenol organic compound 2, 4, 6-TCP is widely used in herbicides, disinfectants, pesticides, papermaking, and plastic industries (Fricker et al., 2014; Jarvinen et al., 1994). It has been detected in various environmental media such as excess sludge, groundwater, surface water, and wastewater (Song et al., 2019). Biological concerns arising from 2,4,6-TCP include genetic toxicity, liver toxicity, immunotoxicity, neurotoxicity, cardiotoxicity, and carcinogenicity (Song et al., 2019; Wang et al., 2020; Yang et al., 2021). It is highly toxic when absorbed by human skin, and can cause obesity, high cholesterol, and high blood pressure in children and adolescents (Yang et al., 2021). Owing to its high toxicity, wide range of pollution, and bioaccumulation properties, 2,4,6-TCP is listed as a priority pollutant by the European Economic Community (EEC) and U.S Environmental Protection Agency (USEPA) (Al-Abduly et al., 2021).

Many technologies have been utilized to remove 2, 4, 6-TCP from wastewater, groundwater, and soil, including adsorption (Saigl, 2020; Tzou et al., 2008), biodegradation (Marsolek et al., 2007), chemical reduction (Liu et al., 2019), and advanced oxidation processes (AOPs) (Karimipour et al., 2021; Xiang et al., 2021b). Owing to the high reactivity, low cost, and non-toxicity of zero-valent iron (ZVI), soil and groundwater remediation has widely adopted the ZVI technology for constructing in-situ permeable reactive barriers (PRBs) (Xiang et al., 2021b). Chlorinated aromatic compounds (2,4,6-TCP, etc.) have been reported to be effectively reduced in the presence of ZVI and the contaminates were converted into small molecular organic substances, which greatly reduces their toxicities (Kim and Carraway, 2000; Liu et al., 2019).

However, ZVI powder can easily undergo agglomeration and passivation in the reaction process. Solid iron corrosion products (e.g. Fe_3O_4 and Fe_2O_3) were generated, which are known to in situ coat the surface of aggregates (Tao et al., 2022; Wang et al., 2020). In particular, in the engineering application of PRBs, ZVI can easily be passivated (reactivity loss) after a period of application (Tao et al., 2022). Some carrier materials have been used to support ZVI powder and improve its performance, including biochar (Liu et al., 2019), graphene oxide (Wang et al., 2022), zeolite (Eljamal et al., 2022), magnesium hydroxide shell (Falyouna et al., 2022), titanium nanowires (Falyouna et al., 2022), etc. The use of iron-supported materials can effectively prevent ZVI agglomeration and passivation (Maamoun et al., 2019). However, solid precipitates such as iron oxide and sulfates could be generated and block the PRBs. The blockage and compaction of powdered and nano-ZVI which are dispersed on supports in PRBs has still not been solved entirely.

Compared with ZVI remediation technology, technology based on microorganisms was also widely adopted for the remediation of contaminated soil and groundwater. The biodegradation and mineralization of 2,4,6-TCP at a lower cost could be achieved in anaerobic environments (Lin et al., 2019). Song et al. revealed that the acclimated microorganisms have a strong ability to dechlorinate and degrade 2,4,6-TCP and phenol ring-openings (Song et al., 2019). However, microorganisms also have the problem of biomass loss and inactivation in the application of the site (Wang et al., 2000). The biodegradation efficiency largely depends on the survival rate of microorganisms in the actual environment. The immobilization carriers were developed to immobilize and protect microorganisms and subsequently enhance the biodegradation rate (Wang et al., 2000). In the construction of in-situ microorganism augmented permeable reactive barriers (Bio-PRBs), zeolite, polyurethane foam, natural loofah sponge (OLS), and other carriers were used to immobilize the microorganisms (Wang et al., 2017). However, the immobilization and remediation performance of Bio-PRBs was still affected by other factors, especially, such as the biological clogging of stuffing (Wang and Wu, 2019b). Furthermore, before PRBs installation, the PRBs dismantling or materials replacement should be considered as part of the design

(Phillips, 2009). Whereas, in the actual field remediation, the maintenance and replacement of reaction materials were not convenient in conventional PRBs because the single-use wall barriers and reaction wells were always constructed (Bekele et al., 2019; Li et al., 2021).

To resolve the aforementioned problems, calcium alginate (CA) embedded ZVI (ZVI @ CA) gel beads and CA embedded biochar (BC) immobilized microorganism (BC&Cell@CA) gel beads were developed. ZVI@CA gel beads avoid the occurence of powdered ZVI hardening in Bio-PRBs. Moreover, BC&Cell@CA gel beads provide large porosity between beads and slow-release microorganism cells, then avoid biological clogging of stuffing in Bio-PRBs. The field-scale modular vertical reaction device which can conveniently replace material also was designed in this study. Their performances were tested by constructing the lab-scale and field-scale BC-immobilized microorganism-augmented ZVI permeable reactive barriers (Bio-PRBs). As a coating agent, CA has the advantages of non-toxicity, harmlessness, simple production, low price, biodegradability, and high mechanical strength (Lemic et al., 2021; Park et al., 2021). The surface of CA is porous and has abundant functional groups, such as hydroxyl, carboxyl, and amino groups, which can uniformly load ZVI and microorganisms on biochar and significantly improve its mechanical stability. Direct contact between organic pollutants and microorganisms also produces toxicity, especially for chlorophenol (Wang et al., 2000). Microorganisms in CA gel beads can effectively avoid direct contact with pollutants. Additionally, as the microorganism carrier, biochar can not only provide attachment sites for microorganisms but also enhance the abundance of microorganisms to form an effective cometabolism-reduction dechlorination system (Wang et al., 2000). In the in-situ remediation process of contaminated soil and groundwater, previous other types of Bio-PRBs systems had shown the advantages of low maintenance costs and no energy consumption (Amoako-Nimako et al., 2021).

In the study, the metabolic mechanism of 2,4,6-TCP and transformation paths of iron oxides in the ZVI@CA, BC&Cell@CA, and BC&Cell@CA+ZVI@CA systems were investigated by the liquid chromatography-mass spectrometry (LC-MS), scanning electron microscopy (SEM)-mapping, X-ray diffraction

(XRD), X-ray photoelectron spectroscopy (XPS), and high-throughput sequencing. The multi-process reaction model of 2,4,6-TCP in different types of Bio-PRBs was constructed, and the transport and transformation process were further investigated by the Hydrus-1D. Furthermore, to further verify the performance of this chemical-biological augmentation technology, the field-scale reaction system was designed to remediate the chlorinated organic compounds and Benzene Toluene Ethylbenzene & Xylene contaminated groundwater in a pesticide factory site.

4.3.2 Materials and methods

(1) Experiment reagent

2,4,6-TCP ($C_6H_3Cl_3O$, AR) was purchased from Shanghai Aladdin Biochemical Technology Co., Ltd. ZVI powder (100 mesh, 98%) and sodium alginate ($(C_5H_7O_4COONa)_n$, AR) were purchased from Shanghai Titan Scientific Co., Ltd. NH_4Cl, KH_2PO_4, $CaCl_2$, $MgSO_4 \cdot 7H_2O$, $NaHCO_3$, $FeSO_4 \cdot 7H_2O$, $CoCl_2 \cdot 6H_2O$, H_3BO_3, $ZnSO_4 \cdot 7H_2O$, $CuSO_4 \cdot 5H_2O$, $NiSO_4 \cdot 6H_2O$, $MnCl_2 \cdot 4H_2O$, $(NH_4)_6Mo_7O_{24} \cdot 4H_2O$ yeast extract, sodium lactate, and acetonitrile analytically pure chemicals were all purchased from Shanghai Sinopharm Chemical Reagent Co. Ltd., China.

(2) Preparation and characterization of amendments

The 500 ℃ peanut shell biochar and domesticated dechlorination microorganisms were referenced in our previous study (Wang et al., 2020). The ZVI@CA gel beads were prepared using the ZVI, sodium alginate, and $CaCl_2$. The BC&Cell@CA gel beads were prepared using the peanut shell biochar, dechlorination microorganism, sodium alginate, and $CaCl_2$.

The BC&Cell@CA and ZVI @ CA gel beads were prepared as follows. Firstly, 0.20 g peanut shell biochar was added into 100 mL microorganism solution (1.91 $g \cdot L^{-1}$), after full mixing, it was incubated in a 25 ℃ incubator for immobilization 48 h. Secondly, after the immobilization process was finished, the biochar immobilized microorganism particles were filtered with 0.45-um filter papers. The sodium alginate and $CaCl_2$ were selected as the embedding medium and crosslinking agent, respectively. Thirdly, the obtained biochar immobilized microorganism particles (or 0.20 g ZVI) were added into a 100-mL sodium alginate solution (20 $g \cdot L^{-1}$). After full mixing, it was dripped by a 5.0-mL injection syringe into the 100-mL $CaCl_2$ solution (20 $g \cdot L^{-1}$) and crosslinked for 2 hours at 25 ℃ (Zhao et al., 2020).

(3) The performance test of amendments and the optimization of one-dimensional Bio-PRBs

a. To obtain the 2,4,6-TCP removal efficiency of two amendments, the Control, ZVI@CA, BC&Cell@CA, and ZVI@CA(50%)+BC&Cell@CA(50%) systems (four reaction systems) were designed in the study. In the ZVI@CA (50%)+BC&Cell@CA(50%) system, 3.0 g of ZVI@CA gel beads (prepared by 0.1 g ZVI) and 3.0 g of BC&Cell@CA gel beads (prepared by 0.1 g biochar) were added into 100 mL sealed anaerobic bottles, which contained 100 mL of the nutrient solution. The formula of the nutrient solution was provided as follows.

The 1-L nutrient solution used to support microorganism cultivation contained 0.32 g NH_4Cl, 0.5 g KH_2PO_4, 0.1 g $CaCl_2$, 0.4 g $MgSO_4 \cdot 7H_2O$, 1.2 g $NaHCO_3$, 1.0 mL 10% yeast extract, 0.72 mL 60% sodium lactate, and 6.7 mL trace-element solution. The trace-element solution is 0.4 g $FeSO_4 \cdot 7H_2O$, 0.32 g $CoCl_2 \cdot 6H_2O$, 0.01 g H_3BO_3, 0.2 g $ZnSO_4 \cdot 7H_2O$, 0.04 g $CuSO_4 \cdot 5H_2O$, 0.2 g $NiSO_4 \cdot 6H_2O$, 0.1 g $MnCl_2 \cdot 4H_2O$, and 0.025 g $(NH_4)_6Mo_7O_{24} \cdot 4H_2O$ dissolved in 1-L Milli-Q water (Ma and Wu, 2007).

The initial concentrations of 2,4,6-TCP and amendment were 6.0 $mg \cdot L^{-1}$ and 2.0 $g \cdot L^{-1}$, respectively. In the BC&Cell@CA system, the amendment was replaced by 6 g of BC&Cell@CA gel beads (prepared by 0.2 g biochar). In the ZVI@CA system, the amendment was replaced by 6 g ZVI@CA gel beads (prepared by 0.2 g ZVI). Other materials and methods were the same as the ZVI@CA(50%)+BC&Cell@CA(50%) system. The pH was 7.0 ± 0.5 and two parallel experiments were conducted. The experiments lasted 96 h and the water samples were filtered through a 0.22-μm membrane for the detection of 2,4,6-TCP and intermediates. No amendment was added in the Control. The detailed reaction conditions in the four types of systems are shown as follows (Table 4.1).

Table 4.1 The reaction conditions in four types of reaction systems.

Protocol	2,4,6-TCP ($mg \cdot L^{-1}$)	ZVI@CA (g)	BC&Cell@CA (g)	Simulated groundwater (mL)
Control	6.0	—	—	100
ZVI@CA	6.0	3		100
BC&Cell@CA	6.0		3	100
ZVI@CA(50%) + BC&Cell@CA(50%)	6.0	1.5	1.5	100

b. To test the practical application performance of the BC&Cell@CA and ZVI@CA gel beads, six types of Bio-PRBs were constructed for the remediation of 2,4,6-TCP contaminated groundwater: the Control, ZVI@CA, BC&Cell@CA, ZVI@CA (10%) + BC&Cell@CA (90%), ZVI@CA (90%) + BC&Cell@CA (10%), and ZVI@CA (50%) + BC&Cell@CA (50%). The schematic of one-dimensional Bio-PRBs and the reaction columns are shown in Figure 4.3.

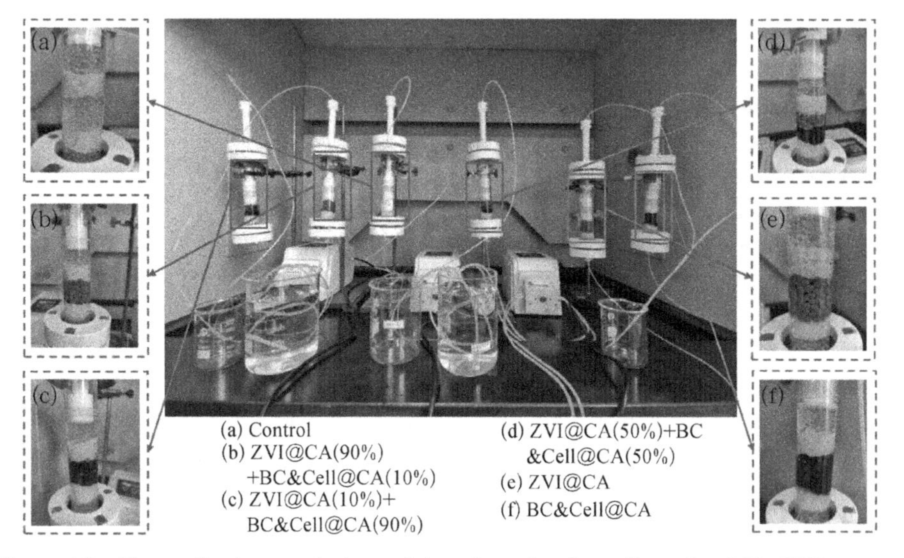

Figure 4.3 The application test device and the schematic of one-dimensional Bio-PRBs; type-1 to type-5 represent the ZVI@CA, BC&Cell@CA, ZVI@CA (50%) +BC&Cell@CA (50%), ZVI@CA (90%) +BC&Cell@CA (10%), and ZVI@CA (10%) +BC&Cell@CA (90%) five types of Bio-PRBs, respectively. Glass beads (Φ 5.0 mm, 11.0 g) were installed in zones ① and ⑥; fine-grain sand (1.57 $g \cdot cm^{-3}$, 70 – 100 mesh, 11.0 g) was installed in zones ② and ⑤; BC&Cell@CA (Φ 5.0 mm, 11.6 g) and/or ZVI@CA (Φ 5.0 mm, 11.6 g) gel bead amendments were installed in zones ③ and ④ in the six types of Bio-PRBs, respectively. No amendment was added in the Control.

The simulated reaction column with a length of 20.0 cm and an inner diameter of 2.5 cm was sealed, and 5 reaction zones were created. The water flow was from zone ① to zone ⑥. Glass beads (Φ 5.0 mm, 11.0 g) were installed in zones ① and ⑥; fine-grain sand (1.57 g · cm^{-3}, 70 – 100 mesh, 11.0 g) was installed in zones ② and ⑤; BC&Cell@CA (Φ 5.0 mm, 11.6 g) and/or ZVI@CA (Φ 5.0 mm, 11.6 g) gel bead amendments were installed in zones ③ and ④ in the six types of Bio-PRBs, respectively. The amendment in the Control was replaced by the glass beads. The detailed packing composition of the six types of Bio-PRBs is provided as follows (Table 4.2).

Table 4.2 The packing composition in six types of Bio-PRBs.

Protocol	2,4,6-TCP (mg · L^{-1})	ZVI@CA (g)	BC& Cell@CA (g)	Simulated groundwater (L)
Control	6.0			1.296
ZVI@CA	6.0	11.6		1.296
BC&Cell@CA	6.0		11.6	1.296
ZVI@CA (10%) + BC&Cell@CA (90%)	6.0	1.2	10.4	1.296
ZVI@CA (90%) + BC&Cell@CA (10%)	6.0	10.4	1.2	1.296
ZVI@CA (50%) + BC&Cell@CA (50%)	6.0	5.8	5.8	1.296

The initial concentration of 2, 4, 6-TCP was 6.0 mg · L^{-1}. To maintain adequate hydraulic retention time for the microorganism growth and the reaction, the velocity was set to 0.3 mL · min^{-1}, and 1.296 L of synthetic 2, 4, 6-TCP contaminated groundwater passed through each Bio-PRBs system using a peristaltic pump for 72 hours (85 pore volumes) (Figure 4.3). Subsequently, the deionized water was pumped into each Bio-PRBs system to wash the residual 2,4,6-TCP. The effluent from the top of each Bio-PRBs system was collected at specific time intervals for obtaining the 2, 4, 6-TCP breakthrough curves. The breakthrough curves were analyzed by the HYDRUS-1D with a two-site sorption coupling chemical and biological reaction model for understanding the reaction process.

c. To obtain the basic model parameters and the Bio-PRBs retardation effect for the ions, KCl was adopted as the tracer to test the performance of Bio-PRBs. The 10 g · L^{-1} KCl solution passed through the six types (Control to type-5) of Bio-PRBs systems (Figure 4.3) at a velocity of 3.0 mL · min^{-1} for 50 min (12 pore volumes). Subsequently, the deionized water was pumped into each Bio-PRBs system to wash the residual KCl. The effluent from the top of each Bio-PRBs system was collected for the evaluation of electrical conductivity. Other materials and methods were the same as experiment (b).

(4) Analyzing the metabolic mechanism and microbial diversity

To obtain the metabolic pathways and reaction mechanism, the intermediates of 2,4,6-TCP were analyzed by LC-MS after the conclusion of experiment (a). Additionally, the solid amendment was collected for SEM, XPS, XRD, and high-throughput sequencing. The change of surface morphology and the secondary mineral species in the interface of the ZVI@CA gel beads were analyzed before and after the reaction. Furthermore, the BC&Cell@CA gel beads (Control, S1, and S2) and free cell solution samples (S3 and S4) were collected from the batch experiment reaction systems after the conclusion of experiment (a) in the above section (3) in this study. Primers 338F (5′-ACTCCTACGGGAGGCAGCAG-3′) and 806R (5′-GGACTACHVGGGTWTCTAAT-3′) were used for the PCR amplification of 16 S rRNA V3-V4 regions in bacteria (Xu et al., 2016). 524F10extF (5′-TGYCAGCCGCCGCGGTAA-3′) and Arch958RmodR (5′-YCCGGCGTTGAVTCCAATT-3′) were used for the PCR amplification of 16 S rRNA V4-V5 regions in archaea (Pires et al., 2012). The microorganism samples were sent to Majorbio (Shanghai, China) for high-throughput sequencing on the MiseqPE300 platform. The data were uploaded to the National Center of Biotechnology Information (NCBI) and the serial number was PRJNA755151.

(5) Multi-process reaction model for 2,4,6-TCP in Bio-PRBs

The breakthrough curves (BTCs) of 2,4,6-TCP in the different types of Bio-PRBs were analyzed by HYDRUS-1D. The reaction process of 2,4,6-TCP was simulated through the two-site sorption coupling chemical and biological reaction model. The water flow parameters were obtained by the Neural Network Predictions in HYDRUS-1D based on the percentages of sand, silt, clay, and bulk density (SSCBD) (Table 4.3).

Table 4.3 The water flow parameters in the Bio-PRBs system, which were obtained by the neural network prediction in the HYDRUS-1D.

Pedotransfer Function	Parameter
Input	
Sand (%)	100
Silt (%)	0
Clay (%)	0
Bulk Density (g · cm^{-3})	1.13
Output	
Qr (-)	0.0524
Qs (-)	0.5013
Alpha (cm^{-1})	0.0456
n (-)	3.0374
Ks (cm · h^{-1})	90.0 *
l (-)	0.5

* The saturation permeability coefficient *Ks* was modified based on the breakthrough curves, the ordinary *Ks* was 45.2267 cm · h^{-1}.

The related reaction parameters for 2,4,6-TCP were inversed based on the observed data in column experiments by the HYDRUS-1D (Tables 4.4 and 4.5).

Table 4.4 The reaction parameters for 2,4,6-TCP in the different types of Bio-PRBs.

Variable	Control	ZVI@CA	BC& Cell@CA	1 : 9[a]	9 : 1	5 : 5
Disp (a/cm)	0.53	0.48	0.35	0.38	0.42	0.20
FRAC (β/-)	70%	40%	20%	20%	10%	20%
KD ($K_d/cm^3 \cdot g^{-1}$)	2.20	10.66	17.65	10.54	16.51	19.00
SNKL1 (λ or k_b/h^{-1})	—	2.40	3.20	2.20	3.00	3.60
ALPHA (α/h^{-1})	0.95	0.92	0.51	0.55	0.75	0.36
R^2	86.78%	95.02%	90.29%	90.13%	80.10%	93.80%

[a] represents the ZVI@CA (10%)+BC&Cell@CA (90%) Bio-PRBs system.

Table 4.5 The reaction parameters for Cl^- and K^+ in different types of Bio-PRBs.

Variable	Control	ZVI@CA	BC& Cell@CA	1 : 9[a]	9 : 1	5 : 5
Disp (a/cm)	0.53	0.48	0.35	0.38	0.42	0.20
FRAC (β/-)	75%	17%	10%	12%	13%	10%
KD ($K_d/cm^3 \cdot g^{-1}$)	0.001	0.23	0.28	1.00	0.30	1.21
SNKL1 (λ or k_b/h^{-1})	—	0.05	0.16	0.15	0.19	0.14
ALPHA (α/h^{-1})	0.005	0.10	0.10	0.04	0.05	0.16
R^2	81.46%	88.79%	79.54%	99.34%	85.55%	92.22%

[a] represents the ZVI@CA (10%) +BC&Cell@CA (90%) Bio-PRBs system.

Subsequently, the calibrated model was used to design the optimal width of Bio-PRBs and to predict the evolution of pollution plumes in different types of Bio-PRBs. The advection-dispersion equation coupled with the two-site sorption and chemical-biological reaction was selected to describe the transport and retention of 2,4,6-TCP in Bio-PRBs. The governing equation is as follows:

$$\frac{\partial C}{\partial t} + f\frac{\rho}{\theta}K_d\frac{\partial C}{\partial t} + (1-f)\frac{\rho}{\theta}\frac{\partial S}{\partial t} = D\frac{\partial^2 C}{\partial z^2} - v\frac{\partial C}{\partial z} + \sum_k (k_{b_k} + \lambda_k)C, \tag{4-1}$$

$$\frac{\rho}{\theta}\frac{\partial S}{\partial t} = \alpha C. \tag{4-2}$$

where C is the 2,4,6-TCP concentration in the aqueous phase ($M \cdot L^{-3}$), f is the fraction of instantaneous retardation to the total retardation, ρ is the medium dry bulk density ($M \cdot L^{-3}$), θ is the moisture content of the pore medium, K_d is the instantaneous equilibrium sorption coefficient of the type-1 site ($L^3 \cdot M^{-1}$), S is the solid phase 2,4,6-TCP concentration ($M \cdot M^{-1}$), D is the hydrodynamic dispersion coefficient [L^2T^{-1}], z is the vertical spatial coordinate [L], and v is the velocity of pore water ($L \cdot T^{-1}$). k_b is the biodegradation rate (T^{-1}), λ is the reaction rate in the chemical process (T^{-1}), and k is the number of reactions. α is the first-order sorption coefficient of the type-2 (kinetic) site (T^{-1}).

(6) Practical application of Bio-PRBs for remediation of pesticide factory groundwater

To test the performance of ZVI@CA (50%) +BC&Cell@CA (50%) Bio-

PRBs for remediation of the actual chlorinated organic compounds and Benzene Toluene Ethylbenzene & Xylene (BTEX) contaminated groundwater in a pesticide factory site, 2.5 kg of BC&Cell@CA gel beads, and 2.5 kg of ZVI@CA gel beads were prepared for constructing the vertical reaction columns. The reaction column with a length of 50.0 cm and an inner diameter of 5.0 cm, the width and spacing of the screen cut are 2 mm and 1.8 cm, respectively. The field-scale device was installed in the monitoring well (Φ 25 cm, depth 12.0 m) of a pesticide factory in Shanghai. The groundwater (1.0 L) was sampled at specific time intervals, and then supplemented with nutrient solution (1.0 L) in the monitoring well for microorganism growth. The experiment lasted 75 days. Groundwater samples were sent to SUEZ Environ. Testing (Shanghai, China) for the detection of chlorinated organic compounds and BTEX.

(7) Analytical methods

The 2,4,6-TCP was detected by high-performance liquid chromatography (HPLC) (Waters 2695, USA). The intermediates of 2,4,6-TCP were analyzed by LC-MS (UPLC 1290; QTOF 6550, Agilent, USA). The surface morphology and elemental mapping were obtained by SEM (SU8010, Hitachi, Japan; GeminiSEM 300, Zeiss, Germany). The elemental analysis was conducted by XPS (EscaLab 250Xi, Thermo Fisher Scientific, USA). The secondary mineral species were analyzed by XRD (model Advance D8, Bruker, Germany). The electrical conductivity was measured by a portable conductivity meter (AZ8362, AZ Instrument, China). The optical density (OD_{600}) was measured by ultraviolet spectrophotometry (Lambda 850, PerkinElmer, USA).

4.3.3 Results and discussion

(1) The 2,4,6-TCP removal kinetics in different batch reaction systems

The results showed that the ZVI@CA+BC&Cell@CA system was the most effective on 2,4,6-TCP removal, and 84% of 2,4,6-TCP removal was achieved in 96 h (Figure 4.4), which was significantly greater than those in the ZVI@CA (47%) and BC&Cell@CA (46%) systems. The removal performance of the ZVI@CA+BC&Cell@CA system increased by 79%. Additionally, the removal of 2,4,6-TCP fitted well with the first-order kinetic model (Wang et al., 2020).

Compared with those in the ZVI@CA (0.004 h^{-1}) and BC&Cell@CA (0.006 h^{-1}) systems, the first-order reaction rate k (0.011 h^{-1}) of 2,4,6-TCP in the combined ZVI@CA+BC&Cell@CA system increased by 175% and 83%, respectively. The chemical-biological augmentation was shown to effectively improve the performance of the combined system, by achieving greater removal efficiency and reaction rate k. The associated reaction mechanism was discussed in the following Section (5) of this study.

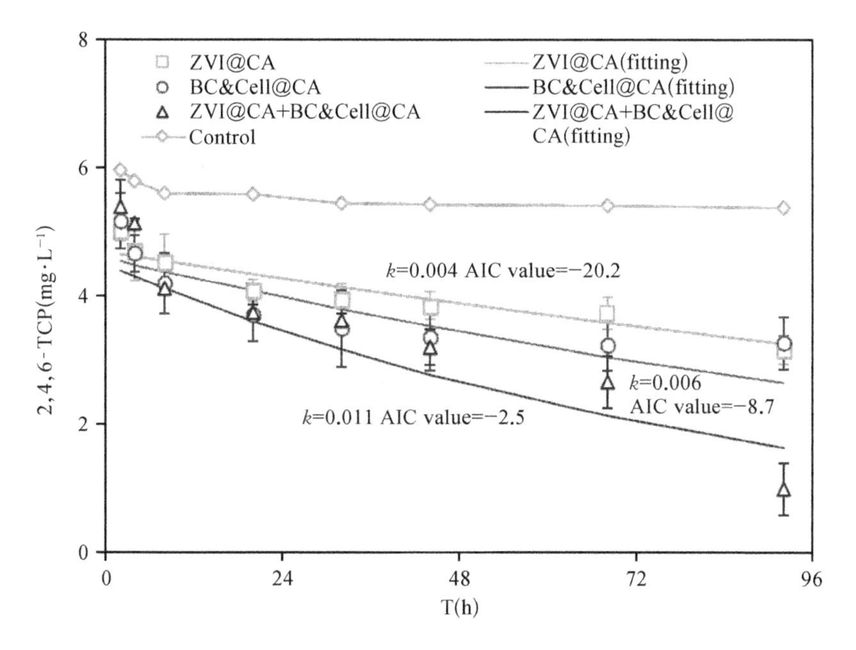

Figure 4. 4 Removal kinetics of 2, 4, 6-TCP in the Control, ZVI @ CA, BC&Cell@CA, and ZVI@CA+ BC&Cell@CA systems. No amendment was added in the Control. AIC was the Akaike's Information Criterion; the lower the value of AIC, the better the fitting of the model with the experimental data (Eljamal, Osama et al., 2022; Maamoun et al., 2021).

(2) The 2,4,6-TCP removal performance of different types of Bio-PRBs

Figure 4.5(a) showed that 82% of 2, 4, 6-TCP breaks through the BC&Cell@CA Bio-PRBs, which was less than those of the ZVI @ CA PRBs (88%) system and Control (96%). Figure 4.5(b) showed that 2,4,6-TCP attenuation of the effluent in 20–85 PVs stage in ZVI @ CA (50%) +

BC&Cell@CA (50%) Bio-PRBs was greater than that in BC&Cell@CA, ZVI@CA (10%) + BC&Cell@CA (90%), and ZVI@CA (90%) + BC&Cell@CA (10%) Bio-PRBs. Furthermore, a significant difference was shown from the sorption and desorption curves in different types of Bio-PRBs in 0 – 20 PVs and 85 – 100 PVs stages. The results demonstrated that the ZVI@CA (50%) + BC&Cell@CA (50%) Bio-PRBs had the best sorption and desorption retardation effects and removal efficiency for 2,4,6-TCP. The change in parameters (Table 4.4) of the reaction model further revealed that the underlying mechanisms were as follows: decrease in the dispersivity a (0.53 to 0.20 cm), and significant increases in the instantaneous equilibrium sorption distribution coefficient K_d (2.2 to 19.00 $cm^3 \cdot mg^{-1}$) and reaction rate λ (2.4 to 3.6 day^{-1}) in the ZVI@CA (50%) + BC&Cell@CA (50%) Bio-PRBs. Additionally, the first-order kinetic sorption coefficient α (0.92 to 0.36 day^{-1}) decreased; however, the fraction β of instantaneous equilibrium sorption decreased (70% to 20%), implying that the fraction of first-order kinetic sorption increased (30% to 80%; Table 4.4).

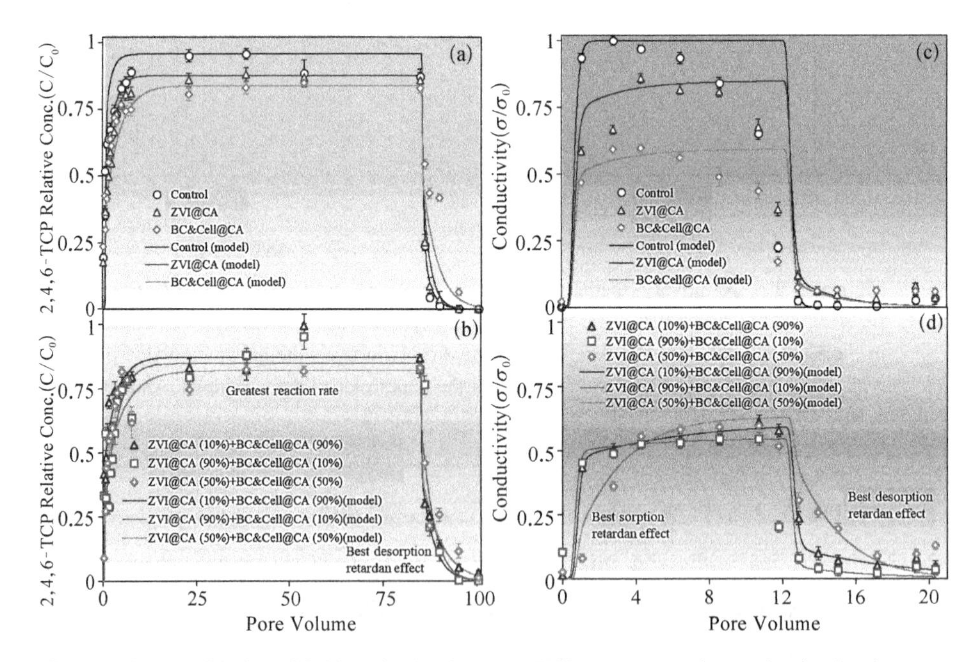

Figure 4.5 (a)-(b) 2,4,6-TCP and (c)-(d) tracer KCl transport and retention in the six types of Bio-PRBs, respectively.

Furthermore, the multi-source data from Cl^- and K^+ breakthrough behaviour was used to adjust and calibrate the parameters of the reaction model (Tables 4.4 and 4.5). The permeability coefficient *K*s (Table 4.3) and dispersivity *a* (Tables 4.4 and 4.5) were verified based on the breakthrough behaviours of 2,4,6-TCP (Figure 4.5 (a)-(b)), Cl^-, and K^+(Figure 4.5 (c)-(d)). The results showed that the instantaneous equilibrium sorption distribution coefficient K_d (0.001 to 1.21 $cm^3 \cdot mg^{-1}$), reaction rate λ (0 to 0.14 day^{-1}), and first-order kinetic sorption coefficient α (0.005 to 0.16 day^{-1}) significantly increased in Bio-PRBs using ZVI@CA (50%) +BC&Cell@CA (50%). Figure 4.5 (c) showed that 58% of Cl^- and K^+ broke through the Bio-PRBs BC&Cell@CA, which was significantly lesser than those in ZVI@CA Bio-PRBs (80%) and Control (99%). The results revealed that the BC&Cell@CA Bio-PRBs had much better performance for preventing Cl^- and K^+ transport compared to the other types of Bio-PRBs. Figure 4.5(d) showed that the conductivity (Cl^- and K^+) attenuation of the effluent in 4 – 12 PVs stage had no significant difference in the BC&Cell@CA, ZVI@CA (50%) +BC&Cell@CA (50%), ZVI@CA (10%) +BC&Cell@CA (90%), and ZVI@CA (90%) +BC&Cell@CA (10%) Bio-PRBs. However, a significant difference was observed from the sorption and desorption curves of Cl^- and K^+in ZVI@CA (50%) +BC&Cell@CA (50%) Bio-PRBs in 0 – 4 PVs and 12 – 20 PVs stages. The above results revealed that the ZVI@CA (50%) + BC&Cell@CA (50%) Bio-PRBs showed the best sorption and desorption retardation effects for Cl^- and K^+, respectively (Figure 4.5 (d)).

Considered together and compared to the other five types of Bio-PRBs, ZVI@CA (50%) +BC&Cell@CA (50%) Bio-PRBs showed the best sorption and desorption retardation effects and removal efficiency for the remediation of Cl^--K^+ type of groundwater contaminated with 2,4,6-TCP.

(3) The change of secondary minerals in the interface of ZVI@CA in the different systems

Compared to the random structure of bare ZVI powder (Figure 4.6 (a)-(d)), the results of the SEM showed that fresh ZVI@CA had a well-developed lamellar structure (Figure 4.7 (a)-(c)). SEM element mapping showed the formation of iron oxides (Figure 4.7 (d)-(g)). The distribution of the Fe was uniform (Figure 4.7 (g)) and O was mainly distributed in the peak and slope of the lamellar structure (Figure 4.7 (f)). Additionally, the results of SEM showed that secondary minerals appeared needle- and club-shaped after the reaction (Figure 4.7 (h)). This was confirmed by the morphology feature of structural Fe(Ⅲ) goethite (α-FeOOH) and the results of XRD (Figure 4.8 (g)). Additionally, the results of SEM (Figure 4.7 (i)) showed that the top of the lamellar structure was severely corroded.

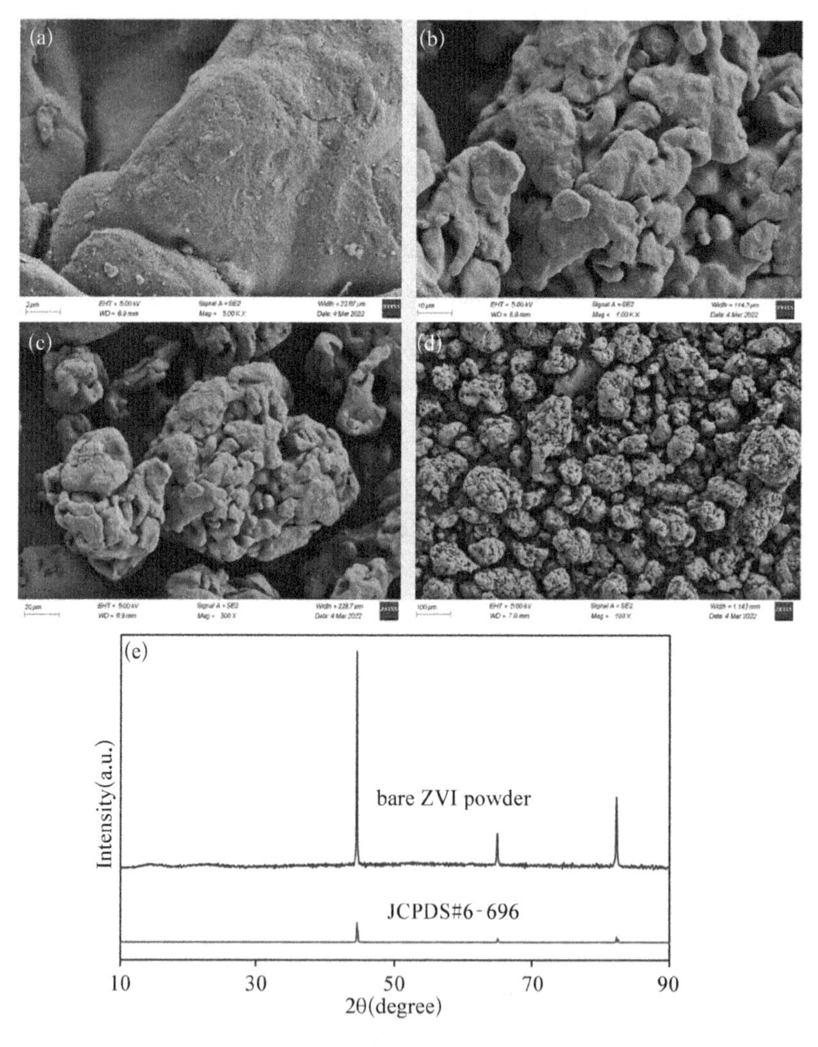

Figure 4.6 (a)-(d) SEM of the fresh bare ZVI powder; (e) XRD patterns of the fresh bare ZVI powder.

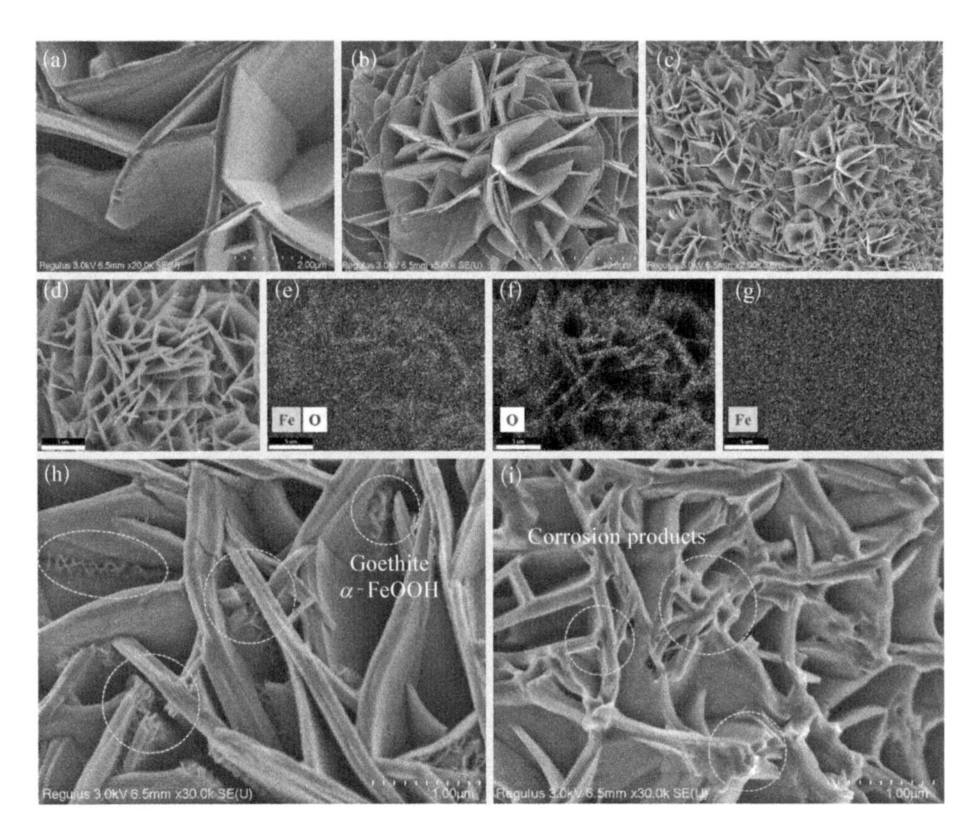

Figure 4.7 (a)-(c) SEM images of the ZVI@CA gel beads before the reaction; (d)-(g) SEM-mapping of fresh ZVI@CA gel beads; (h) SEM images of ZVI@CA gel beads from the ZVI@CA system after the reaction; (i) SEM images of ZVI@CA gel beads from the ZVI@CA + BC&Cell@CA system after the reaction.

Figure 4.8 (a)-(f) showed the XPS spectra of ZVI@CA before and after the reaction in the different systems. The peak of Fe $2p_{3/2}$ at 710.1 eV was attributed to Fe^{2+}. The peak of Fe $2p_{3/2}$ at 711.6 eV and Fe $2p_{1/2}$ at 724.2 eV was attributed to Fe^{3+}(Rajajayavel and Ghoshal, 2015). The results demonstrated the existence of Fe^{2+}(25.84%) and Fe^{3+}(74.16%) oxides in the interface of ZVI@CA before the reaction (Figure 4.8 (a)). Additionally, iron oxide (Fe-O) and iron hydroxide (Fe-OH) were detected in the range of O 1s (Figure 4.8 (b)). The two peaks located at 529.6 and 531.2 eV were attributed to the iron oxide (Fe-O) (31.34%) and iron hydroxide (Fe-OH) (68.66%) species (Kim et al., 2011b). After the reaction, only Fe^{3+} $2p_{3/2}$ and Fe^{3+} $2p_{1/2}$ of iron oxide showed peaks (711.6 and 724.2 eV) (Rajajayavel and Ghoshal, 2015), which demonstrated that only Fe^{3+}

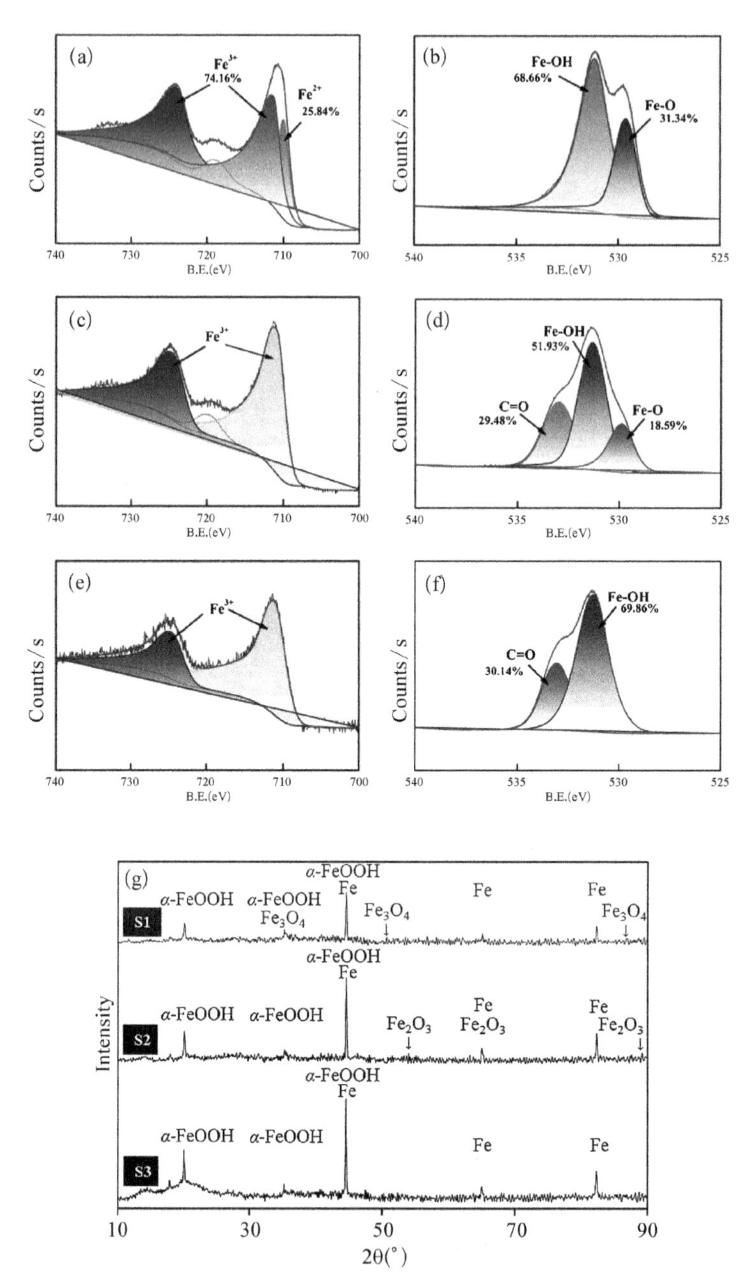

Figure 4.8 XPS spectra of ZVI@CA gel beads in the three systems: (a)-(b) Fe 2p and O 1s regions of ZVI@CA gel beads before the reaction; (c)-(d) Fe 2p and O 1s regions of ZVI@CA gel beads from the ZVI@CA system after the reaction; (e)-(f) Fe 2p and O 1s regions of ZVI@CA gel beads from the ZVI@CA + BC&Cell@CA system after reaction; (g) XRD patterns of the ZVI@CA gel beads before and after the reaction. S1 represented the ZVI@CA gel beads before the reaction. S2 and S3 represented the ZVI@CA gel beads from the ZVI@CA and ZVI@CA + BC&Cell@CA systems after the reaction, respectively.

oxides existed in the interface of ZVI@CA in the ZVI@CA and ZVI@CA+BC&Cell@CA systems after the reaction (Figure 4.8 c and 4.8 e). The three peaks located at 530.1, 531.4, and 532.9 eV were attributed to iron oxide (Fe-O) (18.59%), iron hydroxide (Fe-OH) (51.93%), and C=O (29.48%) species of ZVI@CA in the ZVI@CA system (Figure 4.8 (d)) (Kim et al., 2011b). The two peaks located at 531.2 and 533.0 eV were assigned to iron hydroxide (Fe-OH) (69.86%) and C = O (30.14%) species of ZVI@CA from the ZVI@CA+BC&Cell@CA system (Figure 4.8 (f)). C = O is presumably derived from the intermediate dichloro-2, 6-benzoquinone (DCQ) of 2, 4, 6-TCP, which was confirmed by LC-MS (Figure 4.9 (b)).

The XRD results (Figure 4.8 (g)) confirmed that the iron oxides in the surface of ZVI@CA gel beads mainly comprised goethite α-FeOOH (JCPDS#26-792) and Fe_3O_4(JCPDS#2-1035) before the reaction. This was consistent with the Fe-O (31.34%) and Fe-O-H (68.66%) results of XPS (Figure 4.8 (a)). This was because ZVI@CA gel beads were partially oxidized to α-FeOOH and Fe_3O_4 during their preparation and storage. Subsequently, Fe^{2+} was oxidated to Fe^{3+}(Figures 4.8 (c) and 4.8 (e)) during the removal of 2,4,6-TCP. Fe_3O_4 was removed, and Fe_2O_3 (Figures 4.8 (g)-S2 and 4.8 (g)-S3) was generated. The main secondary minerals in the surface of the ZVI@CA system were α-FeOOH (JCPDS#26-792) and Fe_2O_3 (JCPDS#3-812) (Figure 4.8 (g)-S2). Results of XPS confirmed that the contents of Fe-O and Fe-OH changed to 18.59% and 51.93%, respectively (Figure 4.8 (d)). However, the main secondary mineral was α-FeOOH (69.86%; JCPDS#26-792) (Figures 4.8 (f) and 4.8 (g)-S3) in the ZVI@CA+BC&Cell@CA system. Under the synergistic effect of the ZVI@CA and BC&Cell@CA systems, the Fe^{2+} and secondary iron oxide minerals are more inclined to form goethite α-FeOOH.

(4) The diversity and abundance of bacteria and archaea, and the reaction mechanism

Archaeal and bacterial communities

Figure 4.10(a) showed the archaeal composition at the phylum level. The abundance of *Halobacterota* in S1 gel beads (93.44%) of the BC&Cell@CA system and S2 gel beads (90.06%) of the ZVI@CA+BC&Cell@CA system was significantly greater than that of the Control (87.02%). We believed that some

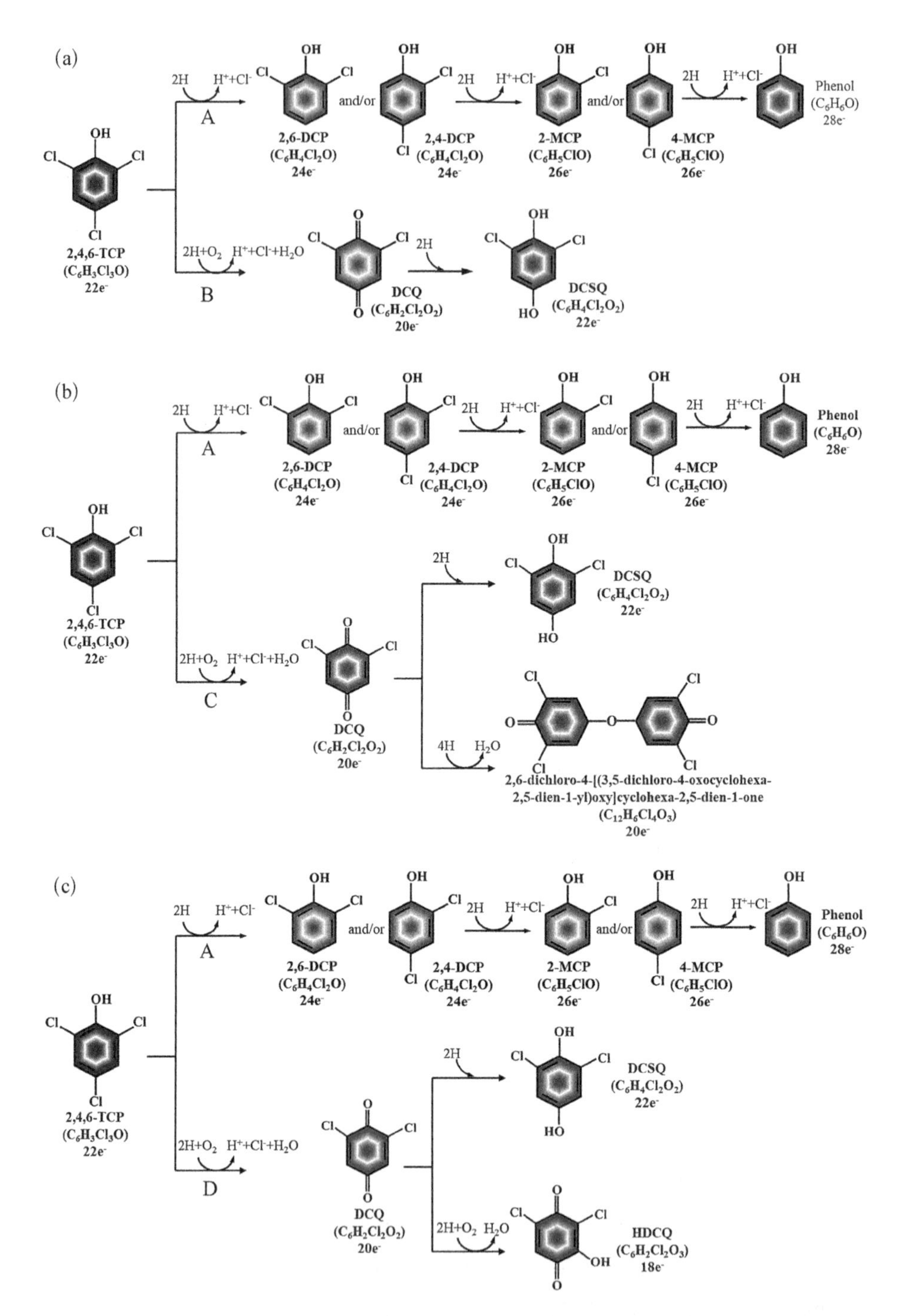

Figure 4.9 Reaction pathways of 2, 4, 6-TCP and intermediates in the (a) ZVI@CA, (b) BC&Cell@CA, and (c) ZVI@CA + BC&Cell@CA systems.

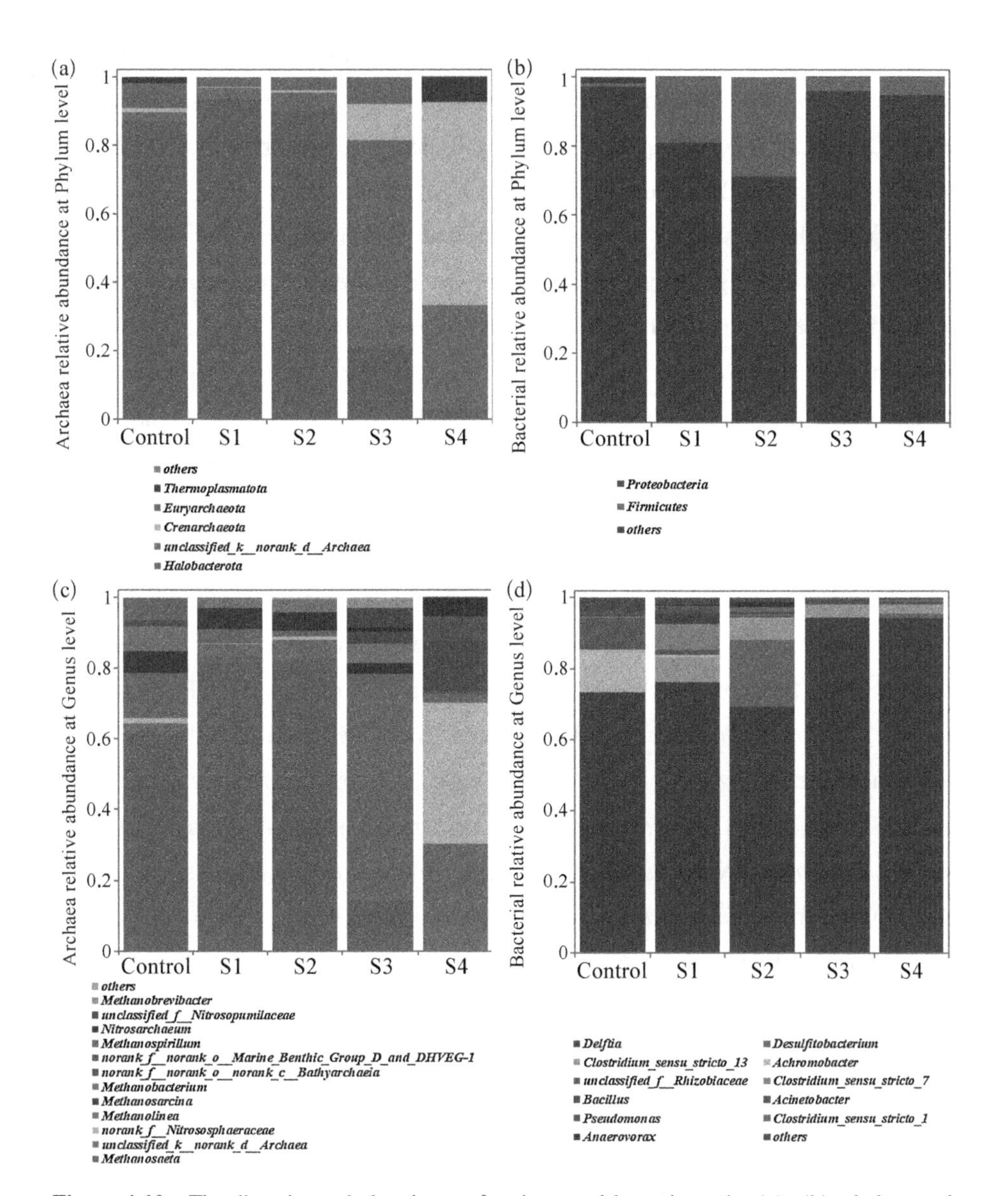

Figure 4.10 The diversity and abundance of archaea and bacteria at the (a)−(b) phylum and (c)−(d) genus levels. Control represented the fresh BC&Cell@CA gel beads sample before they were used; S1 and S2 represented the BC&Cell@CA gel beads samples collected from the BC&Cell@CA and BC&Cell@CA + ZVI@CA systems after 3 days, respectively; S3 and S4 were the solution samples collected from the BC&Cell@CA and ZVI@CA+BC&Cell@CA systems after 3 days, respectively.

dehalogenation archaea (e.g. *Methanosarcina* and *Methanospirillum*) belonged to the phylum *Halobacterota* (Holliger, 1990; Wang and Wu, 2017). Moreover, *Crenarchaeota* (59.37%) and *Thermoplasmatota* (7.10%) in the S4 solution of the ZVI@CA+BC&Cell@CA system were significantly greater than those (10.42%, 0.00%) in the S3 solution of the ZVI@CA system.

Figure 4.10 (c) showed the genus-level archaeal composition. The abundance of *Methanosaeta* in S1 gel beads (83.27%) of the BC&Cell@CA system and S2 gel beads (82.63%) of the ZVI@CA+BC&Cell@CA system was significantly greater than that of the Control (61.88%). The *Methanosaeta*, *Methanolinea*, *Methanosarcina*, *Methanobacterium*, and *Methanospirillum* were found in gel beads. *Methanosarcina* could dechlorinate the chloroethane to ethane (Holliger, 1990). *Methanobacterium* was the cometabolism archaea for the dechlorination of 1,1,1-trichloroethane to 1,1-dichloroethane (Egli et al., 1987). The abundance of *Methanospirillum* increased by 105% under the 1,1,1-trichloroethane condition (Wang and Wu, 2017). The above five strains were presumed to be the dominant cometabolism functional archaea for the dechlorination of 2,4,6-TCP. Moreover, *norank _f_Nitrososphaeraceae* (39.84%) and *Nitrosarchaeum* (5.38%) in the S4 solution of the ZVI@CA+BC&Cell@CA system were significantly greater than those (0.00% and 1.02%, respectively) in the S3 solution of the ZVI@CA system. *Candidatus Nitrocosmicus*, belonging to the *Nitrososphaera* sister cluster, was a novel ammonia-oxidizing archaea species (Wang et al., 2019).

Figure 4.10 (b) showed bacterial compositions at the phylum level. The abundance of *Firmicutes* in S2 gel beads (28.49%) of the ZVI@CA+BC&Cell@CA system was significantly greater than those in S1 gel beads (18.98%) of the BC&Cell@CA system and Control (1.04%). Moreover, *Firmicutes* in the S4 solution (5.26%) of the ZVI@CA+BC&Cell@CA system also was greater than that (4.01%) in the S3 solution of the ZVI@CA system. The abundances of *Proteobacteria* in the S3 and S4 solutions (95.97% and 94.72%) were greater than those (80.96% and 71.20%) in the S1 and S2 gel beads. The dehalogenation bacteria (e.g. *Delftia*, *Desulfitobacterium*, and *Pseudomonas*) belonged to the phylum *Proteobacteria* (Zhang et al., 2018).

Figure 4.10 (d) showed that the dominant bacteria at the genus level in the

ZVI @ CA + BC&Cell@CA system were *Delftia and Desulfitobacterium.* The abundances of *Delftia* in the S3 and S4 solutions (94.27% and 94.07%) were both greater than those in the Control, S1, and S2 gel beads (73.36%, 76.16%, and 69.43%, respectively). The abundances of *Desulfitobacterium* in the S2 gel beads (18.62%) and S4 solution (1.36%) of the ZVI@CA+BC&Cell@CA system were both greater than those of the S1 gel beads (0.03%) and S3 solution (0.00%) in the ZVI @ CA system and Control (0. 08%). However, the abundances of *Pseudomonas* in the S1 gel beads (1.26%) and S3 solution (0.25%) of the ZVI@CA system were both greater than those of the S2 gel beads (0.01%) and S4 solution (0.00%) in the ZVI@CA+BC&Cell@CA system.

At the phylum level, the dominant bacteria *Proteobacteria* were the dominant population involved in dechlorination (Song et al., 2019). *Delftia* was the dominant bacteria belonging to *Proteobacteria.* Many species of *Delftia* lived in anoxic-aerobic environments and could dehalogenate and degrade halogenated aromatic compounds. The genus *Delftia* could also dechlorinate and degrade 2,4,6-TCP and 2, 4-DCP (Zhang et al., 2019). According to reports, cometabolism bacteria *Desulfitobacterium* could induce ortho-dehalogenation of 2,4,6-TCP and had a high capacity for chlorophenol dehalogenation (Breitenstein et al., 2007). *Pseudomonas* could tolerate up to 600 mg · L^{-1} of 2,4-DCP through the meta-pathway degradation of 2,4-DCP (Olaniran et al., 2017). Therefore, *Desulfitobacterium* was presumably the functional bacteria for 2,4,6-TCP removal in the ZVI @ CA + BC&Cell@CA system. Conversely, in the BC&Cell@CA system, *Delftia* and *Pseudomonas* were presumed to be the functional bacteria for 2,4,6-TCP removal.

The richness and diversity of archaea and bacteria

After 96 h of reaction, the richness differences of archaea in gel beads were not significant among the Control (Chao1: 12.000), S1 (Chao1: 12.000), and S2 (Chao1: 11.000) (Table 4.6). However, the archaeal richness in the S4 solution (Chao1: 26.857) of the BC&Cell@CA+ZVI@CA system was significantly greater than those in the S3 solutions (Chao1: 13.000) of the ZVI @ CA system and Control (Chao1: 12.000). Moreover, the archaeal diversity of the S4 solution (Shannon: 1.604, Simpson: 0.256) was significantly greater than those of the S3 solution (Shannon: 0.788, Simpson: 0.684) and Control (Shannon: 1.399, Simpson: 0.410). The archaeal diversity in gel beads decreased. Therefore, the results demonstrated that the combination of BC&Cell@CA and ZVI@CA was more beneficial for improving the diversity and richness of archaea in the reaction solution rather than in gel beads.

Table 4.6 The richness and diversity of archaea and bacteria in and outside the BC&Cell@CA gel beads from the BC&Cell@CA and ZVI@CA+BC&Cell@CA systems.

Microorganism	Reaction system	Sampling condition		Chao1[a]	Shannon[b]	Simpson[c]
Archaea	Control	gel beads	before reaction	12.000	1.399	0.410
	BC&Cell@CA(S1)	gel beads	after reaction	12.000	0.736	0.697
	BC&Cell@CA+ZVI@CA(S2)	gel beads	after reaction	11.000	0.788	0.684
	BC&Cell@CA(S3)	solution	after reaction	13.000	1.483	0.392
	BC&Cell@CA+ZVI@CA(S4)	solution	after reaction	26.857	1.604	0.256
Bacteria	Control	gel beads	before reaction	140.370	1.087	0.560
	BC&Cell@CA(S1)	gel beads	after reaction	75.857	1.035	0.592
	BC&Cell@CA+ZVI@CA(S2)	gel beads	after reaction	105.000	1.055	0.521
	BC&Cell@CA(S3)	solution	after reaction	68.000	0.286	0.890
	BC&Cell@CA+ZVI@CA(S4)	solution	after reaction	25.750	0.327	0.886

[a] Chao1 was the index of microbial richness; greater values indicate greater richness.
[b] Shannon was the index of microbial diversity; greater values indicate greater diversity.
[c] Simpson was the index of microbial diversity; lower values indicate greater diversity.

Table 4.6 also presented that the diversity differences of bacteria in gel beads were not significant among the Control (Shannon: 1.087, Simpson: 0.560), S1 (Shannon: 1.035, Simpson: 0.592), and S2 (Shannon: 1.055, Simpson: 0.521). However, the bacterial richness in S2 gel beads (Chao1: 105.000) of the BC&Cell@CA+ZVI@CA system was significantly greater than that in the S3 gel beads (Chao1: 75.857) of the ZVI@CA system. Moreover, the bacterial diversity of the S4 solution (Shannon: 0.327, Simpson: 0.886) was slightly greater than that of the S3 solution (Shannon: 0.286, Simpson: 0.890). The bacterial richness in the reaction solution decreased. Therefore, the results demonstrated that the combination of BC&Cell@CA and ZVI@CA was more beneficial for improving the bacterial richness in gel beads, and bacterial diversity in the reaction solution.

Considered together, the archaeal richness and bacterial diversity in gel beads were not affected. Compared to the BC&Cell@CA system, the BC&Cell@CA+ZVI@CA system had more advantages in improving the diversity and richness of archaea in the reaction solution, bacterial richness in gel beads, and bacterial diversity in the reaction solution.

(5) Reaction mechanism of 2,4,6-TCP and transformation path of iron oxides in different systems

Figure 4.9 (a) showed the proposed reaction pathways of 2,4,6-TCP in the ZVI@CA system. The pathways (A) and (B) were based on the results of LC-MS and previous reports (Christoforidis et al., 2011; Dorathi and Kandasamy, 2012; Souza et al., 2019). Dorathi and Kandasamy (2012) reported that DCP, CP, and phenol were the 2,4,6-TCP dechlorination products under ZVI reduction (Dorathi and Kandasamy, 2012). Moreover, we demonstrated that the 2,6-dichlorophenol (2,6-DCP), 2,4-dichlorophenol (2,4-DCP), 4-monochlorphenol (4-MCP), and 2-monochlorphenol (2-MCP) were produced through the pathway (A) under ZVI conditions (Souza et al., 2019). In our study, we detected phenol, which was an indicator of such a degradation mechanism. Additionally, the intermediate 2,6-dichloro-p-benzosemiquinone radical (DCSQ) was detected; this was consistent with the report from Christoforidis (2011) stating that dichloro-2,6-benzoquinone (DCQ) and 2,6-dichloro-p-benzosemiquinone radical (DCSQ) were produced through pathway (B) through the deprotonation and elimination of HCl (Christoforidis et al., 2011).

The proposed 2,4,6-TCP reaction pathways in the BC&Cell@CA system is shown in Figure 4.9 (b). Christoforidis et al. (2019) reported that phenol was the final microbial dichlorination product through pathway (A) (Christoforidis et al., 2011). Except for the dechlorination reaction pathway (A), the results of LC-MS indicated the presence of intermediates dichloro-2, 6-benzoquinone (DCQ), 2, 6-dichloro-p-benzosemiquinone radical (DCSQ) and chlorinated dimers 2, 6-dichloro-4-[(3, 5-dichloro-4-oxocyclohexa-2, 5-dien-1-yl) oxy] cyclohexa-2,5-dien-1-one ($C_{12}H_6Cl_4O_3$). The chlorinated dimers were obtained from coupling reactions between DCQ through the pathway (C) (Christoforidis et al., 2011).

The proposed 2,4,6-TCP reaction pathway in the ZVI@CA+BC&Cell@CA system is shown in Figure 4.9 (c). The LC-MS result showed that DCQ, DCSQ, and HDCQ were detected. The 2-hydroxyl-dichloro-2,6-benzoquinone (HDCQ) was a hydroxo derivative of DCQ produced through pathway (D) (Christoforidis et al., 2011). Under the synergistic effect of the ZVI@CA and BC&Cell@CA systems, the chlorinated dimers have not been detected. This implies that the ZVI@CA+BC&Cell@CA could inhibit the coupling reactions between DCQ.

Based on the evidence from the change of microbial diversity and Figure 4.10, the functional bacteria playing a key role in 2,4,6-TCP removal were confirmed to be sulfate-reducing bacteria *Delftia*, *Desulfitobacterium*, and *Pseudomonas* (Christoforidis et al., 2011; Olaniran et al., 2017; Zhang et al., 2019). Furthermore, *Desulfitobacterium* is known to be a very versatile microorganism capable of using a large range of electron acceptors, such as natural or man-made halogenated organic compounds, humic acids, metal sulfites, and nitrate (Villemur et al., 2006). The research revealed that the sulfate-reducing bacteria *Desulfovibrio ferrophilus* produced electron intermediates, which could pass through the sodium alginate microspheres and finally reduce Fe(III) oxides (Liang et al., 2021). Notably, *Desulfitobacterium* could use the poorly crystalline Fe(III) oxide as the electron acceptor (Shelobolina et al., 2003). Therefore, the probable mechanism in the BC&Cell@CA + ZVI @ CA system was the promotion of *Desulfitobacterium* growth (increased from 0.08% to 18.62%) by an iron ion-rich environment (Figure 4.10 (d)) and the faster co-metabolism removal of 2,4,6-TCP (Breitenstein et al., 2007). Electron intermediates produced from *Desulfitobacterium* successfully reduced and transformed the Fe(III) iron oxides (Figure 4.8). However, in the BC&Cell@CA system, *Delftia* and *Pseudomonas* were presumed to be the functional bacteria for 2,4,6-TCP removal. Based on the XPS, XRD, LC-MS, and the high-throughput sequencing results (Figures 4.7–4.10), the underlying mechanism of 2,4,6-TCP removal and the transformation paths of iron oxides in the ZVI @ CA, BC&Cell@CA, and BC&Cell@CA+ZVI@CA systems are shown in Figure 4.11.

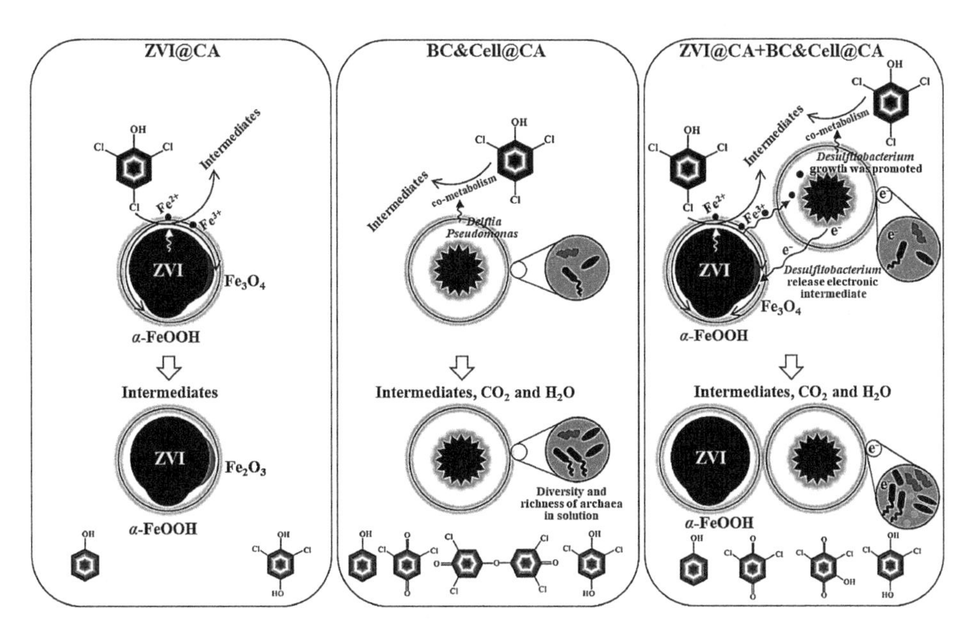

Figure 4.11 Proposed reaction mechanism of 2, 4, 6-TCP and transformation paths of iron oxides in the ZVI@CA, BC&Cell@CA, and BC&Cell@CA+ZVI@CA systems based on our research and previous reports (Breitenstein et al., 2007; Liang et al., 2021; Olaniran et al., 2017; Shelobolina et al., 2003; Zhang et al., 2019).

(6) Optimizing the width of Bio-PRBs through the calibrated multi-process reaction model

The 2,4,6-TCP reaction parameters (Table 4.4) for different types of Bio-PRBs were obtained through the column experiment results (Figure 4.5). The calibrated model was used to design the optimal width of Bio-PRBs and predict the evolution of pollution plumes in different types of Bio-PRBs. Figure 4.12 shows the breakthrough curves of 2,4,6-TCP in observation points of different widths of Bio-PRBs by the model prediction. Figure 4.12 a showed that the optimal width of PRBs with ZVI@CA amendment was 160 cm (containing 35.2 cm width of ZVI@CA and 124.8 cm width of quartz sand), which can achieve at least 90% of 2,4,6-TCP removal. However, that of Bio-PRBs with the ZVI@CA (50%) + BC&Cell@CA (50%) amendment was only 120 cm (containing 13.2 cm width of ZVI@CA, 13.2 cm width of BC&Cell@CA and 93.6 cm width of quartz sand) (Figure 4.12 (a)). The optimal width and cost were significantly less than that of

PRBs with the ZVI@CA amendment. The above results showed that the ZVI@CA (50%) +BC&Cell@CA (50%) type of Bio-PRBs had excellent performance for the remediation of 2,4,6-TCP contaminated groundwater. Furthermore, a combination of chemical and biological augmentation technology could significantly reduce the width of Bio-PRBs and save costs.

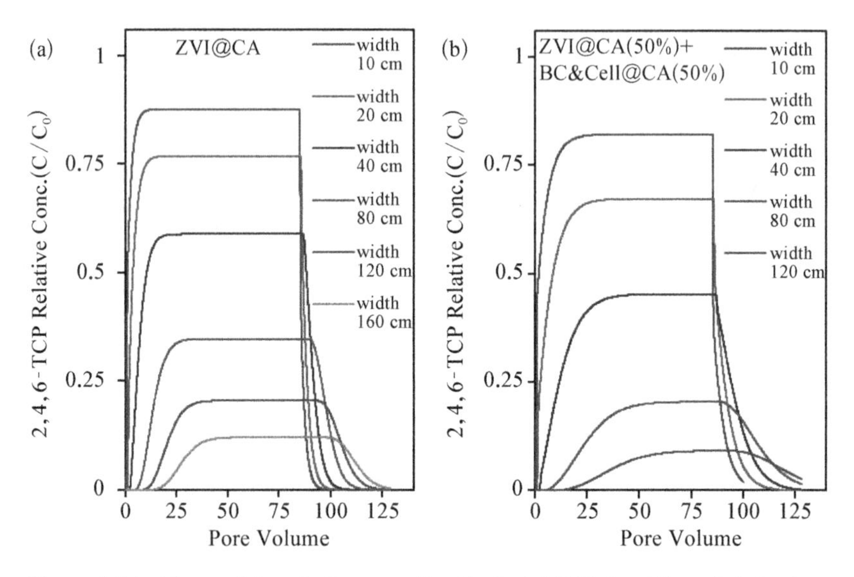

Figure 4.12 The model prediction revealing the 2,4,6-TCP breakthrough behaviour at different widths of Bio-PRBs using (a) ZVI@CA and (b) ZVI@CA (50%) + BC&Cell@CA (50%).

(7) The field test performance of Bio-PRBs for remediation of pesticide factory groundwater

Figure 4.13 showed that the field-scale ZVI@CA (50%) +BC&Cell@CA (50%) reaction columns had excellent performance for in-situ remediation of groundwater contaminated with chlorinated organic compounds and BTEX in a pesticide factory site. The initial average concentrations of the two predominant pollutants Chlorobenzene and Methylbenzene reach 7 322 $\mu g \cdot L^{-1}$ and 8 928 $\mu g \cdot L^{-1}$, respectively (Figure 4.13 (a)-(b)). The removal percentage of chlorinated organic compounds and BTEX (1,2,4-Trichlorobenzene, 1,2-Dichlorobenzene, 1,4-Dichlorobenzene, 2-Chlorotoluene, Bromobenzene, M/p-xylene, Ethylbenzene, Benzene, Methylbenzene, Chlorobenzene) ranged from 61% to 100%. The

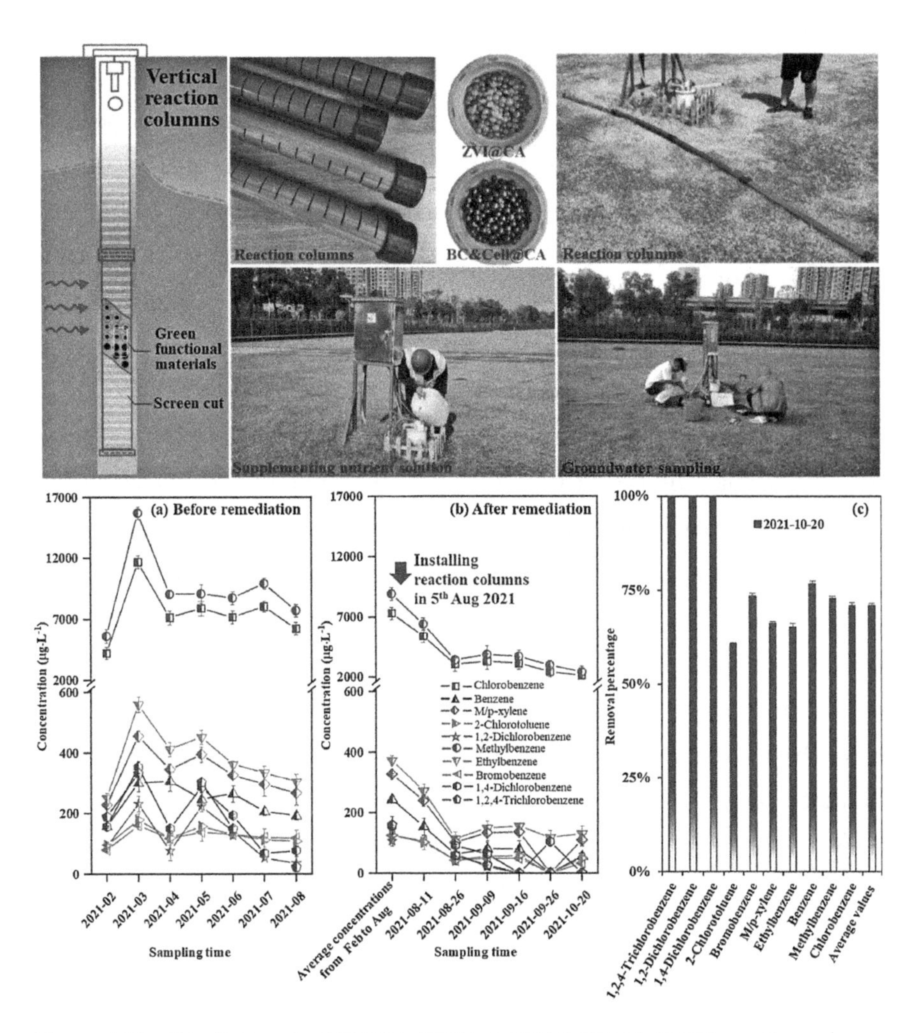

Figure 4.13 The field-scale ZVI@CA (50%) + BC&Cell@CA (50%) reaction columns for in-situ remediation of groundwater contaminated with chlorinated organic compounds and BTEX in a pesticide factory site. The concentrations change of pollutants before remediation during February and August (a), after remediation during August and October (b), and pollutant removal percentages (c).

average removal percentage reached 71% (Figure 4.13 (c)). 1,2,4-Trichlorobenzene (158 $\mu g \cdot L^{-1}$), 1, 2-Dichlorobenzene (106 $\mu g \cdot L^{-1}$), 1, 4-Dichlorobenzene (155 $\mu g \cdot L^{-1}$) were completely removed. After remediation, two high-concentration pollutants Chlorobenzene (7 322 $\mu g \cdot L^{-1}$) and Methylbenzene (8 928 $\mu g \cdot L^{-1}$) decreased to 2 120 and 2 430 $\mu g \cdot L^{-1}$, respectively. The removal percentage reached 71% and 73%, respectively (Figure 4.13 (c)). Moreover, the pentachlorophenol (162 $\mu g \cdot L^{-1}$) was detected on 4th August 2021. After the reaction columns were installed on 5th August 2021, the pentachlorophenol has also not been detected in the following monitoring. The field test results demonstrated that the pollutants were significantly eliminated by field-scale ZVI@CA (50%) + BC&Cell@CA (50%) reaction columns. It is a promising technology for the efficient remediation of actual groundwater contaminated with chlorinated organic compounds and BTEX.

(8) Conclusions of the case study

Taken together, the study explored the effect of chemical-biological augmentation on the performance of Bio-PRBs for 2,4,6-TCP removal. The ZVI@CA and BC&Cell@CA gel beads were developed to construct different types of Bio-PRBs; the underlying mechanism and multi-process reaction model of 2,4,6-TCP in Bio-PRBs was investigated. The reaction rate of 2,4,6-TCP increased by 175% in the combined chemical-biological system. Two-site sorption coupling chemical-biological multi-process reaction model can be used for revealing 2,4,6-TCP transport behaviour and optimizing Bio-PRBs. Chemical-biological augmentation significantly improved the retardation effect of Bio-PRBs for 2, 4, 6-TCP. Moreover, the versatile functional bacteria *Desulfitobacterium* was crucial in the transformation of Fe (III) iron oxides. The diversity and richness of archaea in the reaction solution were improved by ZVI@CA gel beads addition. Furthermore, field test results demonstrated it is a promising technology to construct vertical reaction columns or horizontal Bio-PRBs (Figure 4.14) for the efficient remediation of actually contaminated groundwater.

4.4 Conclusions and suggestions

Microorganisms promote the liquefaction of metal minerals into metal ions, which improves the abundance of functional microorganisms, thus forming a positive cycle by microbe-mineral interaction. Moreover, in the coupling of microorganisms with non-metallic packing type Bio-PRBs, the k promotion is controlled by synergistic effects of biodegradation, biological reduction, and electron donor groups.

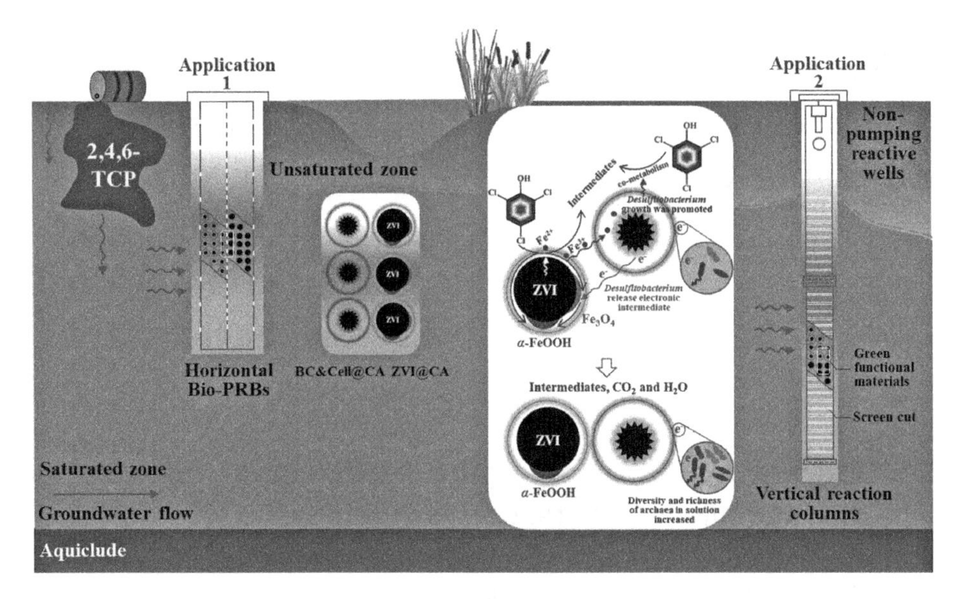

Figure 4. 14 Vertical reaction columns or horizontal Bio-PRBs for the remediation of contaminated groundwater.

The above research has demonstrated that microbe-mineral interaction Bio-PRBs are effective, however, some matters need attention. Suggestions on the application and development of Bio-PRBs are proposed.

The microbe-mineral interaction in novel Bio-PRBs is more complicated than expected, and the interaction between microorganisms and chemical media remains to be further explored on a scale. Especially, the interaction process of microorganisms, minerals, and various other elements (Fe, S, CaO_2, etc.) in the geochemical reaction of real groundwater.

Chapter 5 Microbe-mineral Interaction for Eliminating Emerging Contaminant

5.1 Current status of emerging contaminant removal by microbe-mineral interaction technology

Emerging contaminants (ECs), such as pharmaceuticals and personal care products (PPCPs), pharmaceuticals (PhACs), and pesticides have been widely detected in the environmental system (Birch et al., 2015; Gogoi et al., 2018). These contaminants may threaten human health via the food chain (Sengupta et al., 2021). PPCPs have persistence, bioaccumulation, and toxicity; mainly enter the water environment through landfill leachate and sewage treatment plants, etc. (Zhang et al., 2017). The abuse of antibiotics threatens human health (Srain et al., 2021). Studies have shown that Bio-PRBs can efficiently remove TC (Huang et al., 2017), PhACs (Vijayanandan et al., 2018), and 1,4-dioxane (Yang et al., 2018) from soil and groundwater.

Huang et al. adopted anaerobic activated sludge with ZVI, ZVI only, and anaerobic activated sludge with silica, respectively, to construct three Bio-PRBs columns for remediating simulated TC-contaminated groundwater (Figure 5.1) (Huang et al., 2017). Compared to the ZVI-PRBs (40%) or sludge Bio-PRBs (10%), the η (50%) of TC increased by 25%–400% in sludge+ ZVI Bio-PRBs; and the k (0.021 d^{-1}) of TC increased by 40%–556% in sludge+ ZVI Bio-PRBs than that of ZVI-PRBs (0.015 d^{-1}) and sludge Bio-PRBs (0.0032 d^{-1}). Additionally, TC as a carbon source was consumed by microorganisms and produced intermediates in sludge Bio-PRBs, but the biodegradation ability was relatively limited (Figure 5.1). Whereas TC was removed through ZVI adsorption, chemical degradation of ZVI, and biodegradation in sludge + ZVI Bio-PRBs; TC was partially degraded to intermediates, and a small part of TC was completely degraded to CO_2 and H_2O. Moreover, EPS produced by microorganisms could also remove some of TC, and ZVI stimulated microorganisms to secrete more EPS. Fe^{2+} and Fe^{3+} were essential elements for microbial growth, and microorganisms promoted electron transfer of ZVI, thus forming a good cycle of biological co-metabolism and ZVI reduction (Figure 5.1) (Huang et al., 2017).

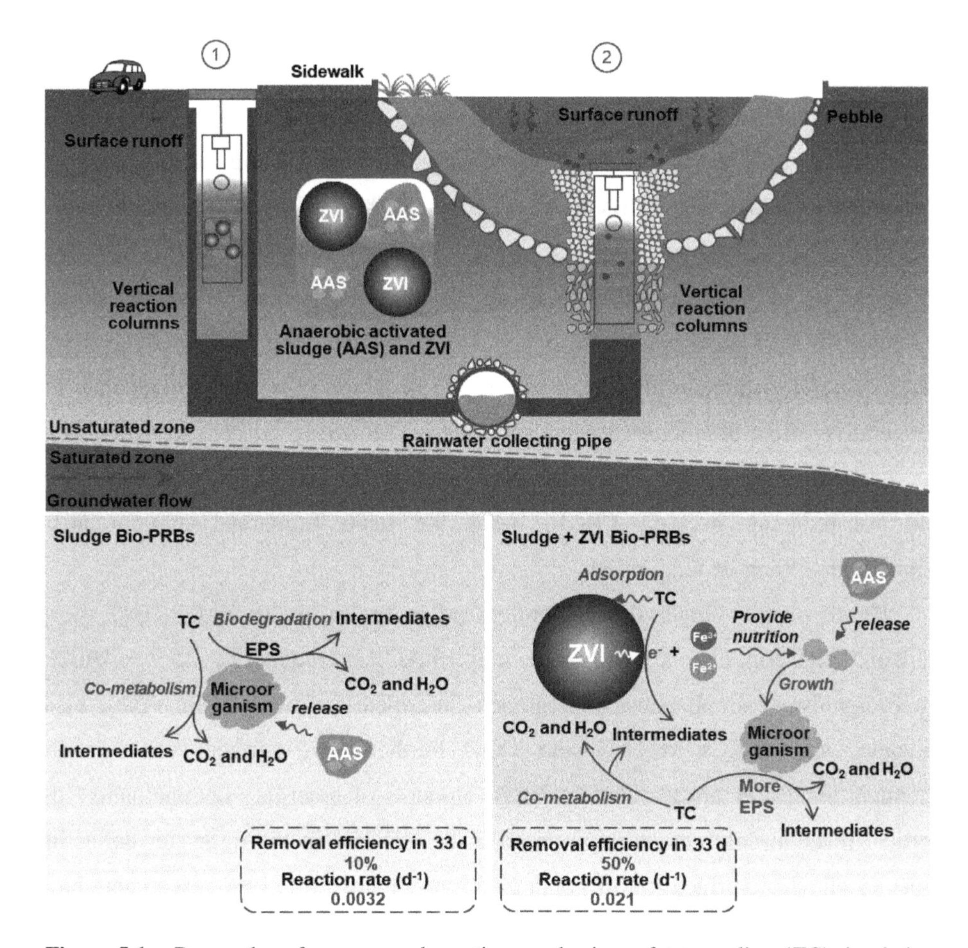

Figure 5.1 Removal performance and reaction mechanism of tetracycline (TC) in sludge Bio-PRBs; and sludge + ZVI Bio-PRBs (Huang et al., 2017).

Vijayanandan et al. adopted clay composite adsorbent immobilized biofilm with glucose (type 1), or only clay composite adsorbent (type 2), or sand and organic matter (Control) to construct three pilot-scale Bio-PRBs for treating simulated atenolol and ciprofloxacin contaminated groundwater. Glucose was as an organic source to maintain microorganism activity (Vijayanandan et al., 2018). Compared to the Control (0.25 d^{-1} and 0.17 d^{-1}), the k (0.34 d^{-1} and 0.39 d^{-1}) of type 1 Bio-PRBs for atenolol and ciprofloxacin increased by 36% - 129%. The coupling of adsorption and biodegradation significantly removed atenolol and ciprofloxacin through more adsorption sites and degrading bacteria. Additionally, the performance of Bio-PRBs was enhanced with the increase in barrier thickness and the decrease in influent flow rate (Vijayanandan et al., 2018).

Yang et al. adopted ZVI coupled with sludge to construct a Bio-PRBs column for treating 1, 4-dioxane-contaminated groundwater (Yang et al., 2018). Compared to sludge Bio-PRBs in 21 - 210 d (0.0047 d^{-1}), the removal rate k (0.029 d^{-1}) of 1,4-dioxane increased by 517% in the coupled sludge + ZVI Bio-PRBs in 211 - 300 d, respectively. Biodegradation of 1,4-dioxane was effectively enhanced by ZVI, and compared to 1,1,1-TCA (1.20), the strengthening factors Q (2.20) of 1,4-dioxane increased by 183%, in which Q is used to calculate the cooperative effect of microorganism and ZVI on removing co-contamination. The intermediate products of 1,4-dioxane mainly were CO_2 and H_2O. Moreover, SRB and iron-reducing bacteria (IRB) were the main functional microorganism populations (Yang et al., 2018).

Based on the views above, compared to the conventional PRBs, the η of Bio-PRBs for TC, 1,4-dioxane and PhACs increased by 25% - 400%. Microorganisms simultaneously degraded intermediate products of TC, 1, 4-dioxane, and PhACs to H_2O and CO_2. Furthermore, surface runoff in the chemical plant area always contains large amounts of emerging contaminants, the vertical reaction columns were proposed to be installed in the rainwater collecting pipes to alleviate surface runoff contamination (Figure 5.1). Few researchers have studied the retention and removal of microplastics by Bio-PRBs, which as the potential carrier for per- and polyfluoroalkyl substances (PFAS), PPCPs, PAHs, etc. probably increase the transport risk. Therefore, using Bio-PRBs to remove emerging microplastic combined contaminants will be a meaningful research direction.

5.2 Case study for PFOA elimination by microbe-Fe^0 interaction

5.2.1 Research background

Perfluorooctanoic acid (PFOA), an emerging contaminant, has gained public attention as one of the extensively manufactured per- and polyfluoroalkyl substances (PFAS) across various industrial and consumer applications (Fan et al., 2023; Teaf et al., 2019). Its widespread production and utilization has contributed to its status as a priority environmental concern (Evich et al., 2022; Wee and Aris, 2023). It has found extensive use as surfactant (Ji et al., 2021), fire retardant (Xiao et al., 2017), and textile coatings (Poothong et al., 2019). High stability of the C-F bond (127 kcal·mol^{-1}) renders PFOA resistant to chemical degradation, making it practically recalcitrant and considered undegradable (Joudan and Lundgren, 2022; Lee et al., 2016; Tran et al., 2022). However, the environmental persistence and bioaccumulation of PFOA have raised concerns about its adverse impacts on human health, including endocrine disorders, immune system dysfunction, and cardiovascular disease (Han et al., 2018; Trowbridge et al., 2020). PFOA has been ubiquitously detected in drinking water (Harrad et al., 2019), groundwater (Johnson, 2022), and surface runoff (Tran et al., 2022). Given the potential risks associated with PFOA-contaminated surface runoff and groundwater, it is imperative to develop an environment-friendly and efficient remediation method.

Various technologies have been utilized to eliminate PFOA contamination from wastewater, groundwater, and surface runoff, including adsorption (Sun et al., 2016), membrane filtration (Bulusu et al., 2020), biotransformation (Chiriac et al., 2023), chemical reduction (Liu et al., 2021), and advanced oxidation processes (AOPs) (Pensini et al., 2019). Among these, biotransformation and zero-valent iron (ZVI)-based technology stand out as the most promising green and low-carbon technologies. The excellent remediation efficiency of ZVI and low energy consumption associated with microbial remediation make them highly attractive options (Kumari et al., 2018; Zhu et al.,

2022). ZVI was commonly used in permeable reactive barriers (PRBs) to remove a wide range of contaminants (e.g. heavy metals (Cr (Li et al., 2019), Cu (Yang et al., 2018), and Cd (Eglal and Ramamurthy, 2015), chlorinated organics (TCA (Wang and Wu, 2019a), 2,4,6-TCP (Wang et al., 2022), and 2,4-DCP (Wang et al., 2020))) from surface runoff and groundwater. Moreover, PFOA has been reported to be reduced and partially defluorinated using ZVI evidenced by F^- release, which indicated the potential of ZVI for reductive defluorination (Liu et al., 2020). Microbial remediation presents another effective option for PFOA removal. Yi et al successfully isolated a PFOA-degrading strain YAB1 through domestication and enrichment from soil near a PFAS-production plant. After 96 h of cultivation, YAB1 achieved a degradation efficiency of 32.4% for PFOA (Yi et al., 2016). Similarly, Chiriac et al investigated PFOA degradation by individual bacterial strains. *Pseudomonas aeruginosa* transformed 27.9% of PFOA, while *Pseudomonas putida* converted 19.0% of PFOA after 96 h of incubation (Chiriac et al., 2023).

However, pristine ZVI encounters challenges such as aggregation (Mwamulima et al., 2017), corrosion (Yoon et al., 2011), and passivation (Ansaf et al., 2016), which impede its effectiveness significantly. Consequently, researchers have endeavored to enhance ZVI efficiency in PFOA removal through noble metal doping (Pd (Lawal Wasiu and Choi, 2018)), the immobilization of ZVI on porous materials (biochar (Yang et al., 2018)), and the utilization of common oxidants (persulfate (Parenky et al., 2020)), which are costly in the long run unfortunately. Similarly, microbes without substrate or coatings are vulnerable to environmental factors, posing challenges in achieving sustained remediation process (Narayanan et al., 2023).

In field-scale applications, using ZVI technology or microbial remediation alone has the disadvantages of being time-consuming and vulnerable to changes in environmental factors (Gillespie and Philp, 2013; Lawrinenko et al., 2023). By using two methods in tandem, ZVI stimulates certain bacteria by creating a suitable environment by providing electron donors (H_2(Liu and Lowry, 2006)) or carbon sources (surface modifiers (Xiu et al., 2010)). It also eliminates harmful substances that inhibit bacteria growth. This positive interaction between ZVI and microbes synergistically removes contaminants (Xie et al., 2017). Aged ZVI can be used by iron-reducing bacteria, converting Fe(III) to Fe(II) and further aiding in contaminant reduction (Gerlach et al., 2000).

Here, two CA-embedded amendments were developed to investigate the enhanced removal of PFOA by combining ZVI and microbial remediation. However, current research still lacks a comprehensive understanding of the metabolic pathways and underlying mechanisms involved in microbial remediation of PFOA. Additionally, the interaction mechanisms between minerals and microorganisms that govern PFOA removal remain insufficiently explored.

Hence, this study presented the green cycling strategies for microbe interaction with Fe^0 for PFOA elimination. Functional genes, corresponding enzymes, related key metabolic pathways, and underlying interaction mechanisms were systematically investigated in this study. Specifically, the objectives of this work are: 1) to evaluate the performance and feasibility of employing CZ and CB amendments for PFOA elimination in microbe-Fe^0 interaction systems; 2) to explore PFOA reaction pathways and mineral transformations and elucidate underlying mechanisms by surface morphology and intermediates analysis; 3) to identify functional genes, corresponding enzymes, and related key metabolic pathway that govern PFOA degradation using metagenomics approaches; 4) to stimulate and investigate PFOA transport and reaction behaviors by developing a multi-process coupling model for PFOA based on microbe-Fe^0 interaction in a dynamic vertical reaction column system.

5.2.2 Materials and methods

(1) Chemicals

Sodium alginate ($(C_5H_7O_4COONa)_n$, analytical grade) and Fe^0 powder (100 mesh, 98%) were obtained from Shanghai Titan Scientific Co., Ltd. PFOA ($F\text{-}(CF_2)_7\text{-}COOH$, analytical grade) was obtained from Shanghai Bide Pharmatech Ltd. Other reagents were obtained from Shanghai Sinopharm Chemical Reagent Co. Ltd., China.

(2) Construction of microbe-Fe^0 interaction system

The utilization of peanut shell biochar (500 ℃) and microorganisms was previously reported in our study (Wang et al., 2020). These microorganisms, specifically capable of biodegrading 2,4-DCP, were selectively cultured using PFOA in this investigation, as detailed as follows.

Microorganisms, capable of biodegrading 2,4-DCP was sampled (500 mL) and inoculated in 2 200-mL filtering flasks with 200-mL headspace under not strictly anaerobic conditions. The flasks were sealed with Teflon-lined rubber septa with a pore (Φ1.0 mm) and stored in an incubator at a constant temperature of 25 ℃. Every week, 1 000 mL of the culture solution was removed from the flasks, and the same volume of fresh nutrient solution was added to each flask, and 20.0 mL PFOA solution (500 $mg \cdot L^{-1}$) was injected into each flask. Two months later, the domestication was finished and the flasks were kept in the dark at 25 ℃. Once a month, 1 000 mL liquid from each culture flask was replaced with a fresh nutrient solution of the same volume to replenish the substrate nutrition, and 20.0 mL PFOA solution was injected.

The CZ and CB amendments were prepared to construct the alone Fe^0, alone microbe, and microbe-Fe^0 interaction batch and vertical reaction column systems. The method was provided as follows.

Firstly, 0.20 g peanut shell biochar was added into 100 mL microorganism solution (1.91 $g \cdot L^{-1}$), after full mixing, it was incubated in a 25 ℃ incubator for immobilization 48 h. Secondly, after the immobilization process was finished, the biochar-immobilized microorganism particles were filtered with 0.45-μm filter papers. The sodium alginate and $CaCl_2$ were selected as the embedding medium and crosslinking agent, respectively. Thirdly, the obtained biochar immobilized microorganism particles (or 0.20 g ZVI) were added into a 100-mL sodium alginate solution (20 $g \cdot L^{-1}$). After full mixing, it was dripped by a 5.0-mL injection syringe into the 100-mL $CaCl_2$ solution (20 $g \cdot L^{-1}$), and crosslinked for 2 h at 25 ℃ (Zhao et al., 2020).

In the microbe-Fe^0 interaction batch system, 6.0 g of CZ amendment (using 0.2 g of Fe^0) and 6.0 g of CB amendment (using 0.2 g of biochar) were added to sealed anaerobic bottles containing nutrient solution (100 mL). The initial concentration of PFOA was 1.0 $mg \cdot L^{-1}$. In the alone Fe^0 batch system, the CZ+CB was replaced with CZ (6.0 g). In the alone microbe system, the CZ+CB was replaced with CB (6.0 g) (Table 5.1). All other materials and methods were performed following the microbe-Fe^0 interaction batch system. The composition of the nutrient solution is as follows.

Table 5.1 The reaction conditions in four types of batch reaction systems.

Protocol	PFOA ($mg \cdot L^{-1}$)	CZ (g)	CB (g)	Simulated groundwater (mL)
Control	1.0	—	—	100.0
alone Fe^0	1.0	6.0	—	100.0
alone microbe	1.0	—	6.0	100.0
Microbe-Fe^0 interaction	1.0	6.0	6.0	100.0

The 1-L nutrient solution used to support microorganism cultivation contained 0.32 g NH_4Cl, 0.5 g KH_2PO_4, 0.1 g $CaCl_2$, 0.4 g $MgSO_4 \cdot 7H_2O$, 1.2 g $NaHCO_3$, 1.0 mL 10% yeast extract, 0.72 mL 60% sodium lactate, and 6.7 mL trace-element solution. The trace-element solution is 0.4 g $FeSO_4 \cdot 7H_2O$, 0.32 g $CoCl_2 \cdot 6H_2O$, 0.01 g H_3BO_3, 0.2 g $ZnSO_4 \cdot 7H_2O$, 0.04 g $CuSO_4 \cdot 5H_2O$, 0.2 g $NiSO_4 \cdot 6H_2O$, 0.1 g $MnCl_2 \cdot 4H_2O$, and 0.025 g $(NH_4)_6Mo_7O_{24} \cdot 4H_2O$ dissolved in 1 L ultrapure water (Ma and Wu, 2007).

The microbe-Fe^0 interaction vertical column system consisted of a one-dimensional reaction column (Figure 5.2), CZ, and CB amendments. The column (Φ2.5 cm, 20.0 cm) was sealed and divided into three reaction zones. Coarse sand (density 2.11 g · cm-3, 11.0 g, 10 – 18 mesh) was filled in sections ② and ⑤. CB (Φ5.0 mm, 11.6 g) and/or CZ (Φ5.0 mm, 11.6 g) were filled in sections ③ and ④, respectively. Glass beads (Φ5.0 mm, 11.0 g) were filled in sections ① and ⑥. In the Control group, CZ and CB were all replaced by glass beads. Detailed information on the filling material of the reaction column was provided in Table 5.2.

(3) Reaction and transport behavior of PFOA in batch and vertical reaction column systems

a. To determine the PFOA removal efficiency, the Control, alone Fe^0, alone microbe, and microbe-Fe^0 interaction batch systems were designed (see above Section (2)). The two parallel experiments were conducted and lasted 8 days.

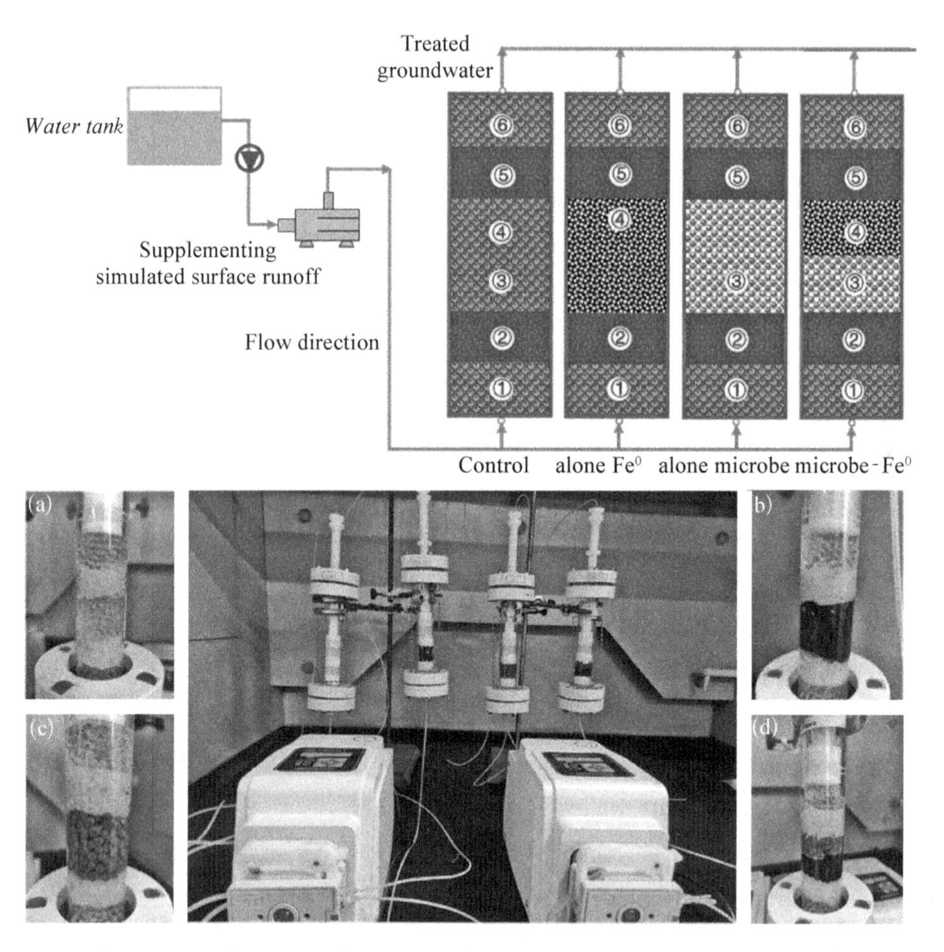

Figure 5.2 The application test device and the schematic of vertical reaction column system. Glass beads (Φ 5.0 mm, 11.0 g) were installed in zones ① and ⑥; coarse sand (2.11 g · cm^{-3}, 11.0 g, 10 – 18 mesh) was installed in zones ② and ⑤; CB(Φ 5.0 mm, 11.6 g) and/or CZ(Φ 5.0 mm, 11.6 g) were installed in zones ③ and ④, respectively. No amendment was added to the Control.

Table 5.2 The packing composition in four types of vertical reaction column systems.

Protocol	PFOA ($mg \cdot L^{-1}$)	CZ (g)	CB (g)	Simulated groundwater (mL)
Control *	1.0			0.864
alone Fe^0	1.0	11.6		0.864
alone microbe	1.0		11.6	0.864
Microbe-Fe^0 interaction	1.0	5.8	5.8	0.864

* In the control group, glass beads replace the CZ and CB amendments.

The water samples were collected and filtered (0.22-μm membrane) for the analysis of PFOA and byproducts.

b. To assess the actual performance of the CZ and CB amendments, Control, alone Fe^0, alone microbe, and microbe-Fe^0 four types of vertical reaction columns were constructed (see above Section (2)) for treating PFOA. The initial C_{PFOA} was 1.0 mg·L^{-1}. Each column was subjected to a velocity of 0.20 mL·min^{-1} and lasted 72 h, and a total of 0.864 L (99 pore volumes) of synthetic PFOA-contaminated water passed through each column (Figure 5.2).

Then the ultrapure water was used to wash out the remaining PFOA. The effluent PFOA was determined for characterizing breakthrough curves. Moreover, the specific reaction columns were used to investigate the 2D-spatial distribution of PFOA in solid media with residual water. It lasted for 72 hours, and no elution process was performed, following the same procedures as the previous reaction column experiments. When the experiments were finished, the column outlet was opened to drain the flowing water. The solid media with residual water were sampled. The distribution of sampling sites is provided in Figure 5.3. Column experiments were performed in duplicates.

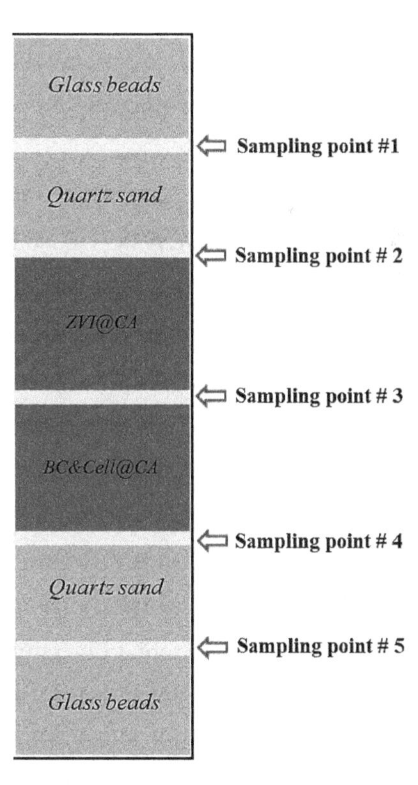

Figure 5.3 The distribution of sampling sites.

Sample pretreatment method: the collected glass beads or quartz sand sample (1.0 – 2.0 g) from each observation point was added to 20 mL of pure methyl alcohol. The slurry mixture was treated with an ultrasonic oscillator (model 5 200 DA, KUNSHAN ULTRASONIC, China) for 30 min, then standing placed for 4 h. The collected CZ and CB sample (1.0 –2.0 g) from each observation point first was added to 20 mL of sodium citrate solution (0.2 mol · L^{-1}) and treated with an ultrasonic oscillator for 2 h. Then 20 mL of pure methyl alcohol was added, and the slurry mixture was treated with an ultrasonic oscillator for 30 min, then standing placed for 4 h. The supernatant was filtered through a 0.22-μm hydrophobic membrane filter for PFOA detection. The obtained data were processed by the *Kriging* interpolation in Surfer 10 (https://www.goldensoftware.com/products/surfer) (Pesquer et al., 2011).

c. To further reveal the underlying transport and transformation processes in the microbe-Fe^0 interaction column system, a microbe-Fe^0 interaction multi-process coupled model was constructed to analyze the breakthrough curves of PFOA. To calibrate the essential model parameters (e.g. dispersion *a* and permeability coefficients *K*s), a tracer study using KCl was conducted. KCl solution (10 $g \cdot L^{-1}$) was injected into the above four reaction column systems. The injection rate was set at 3.0 $mL \cdot min^{-1}$ and lasted 50 min (12 pore volumes). Then ultrapure water was flushed through reaction columns to remove any remaining KCl. The effluent Cl^- concentration was determined. The experimental setup and procedures remained consistent with those described in experiment (b).

(4) Analysis of microbe-Fe^0 interaction mechanism and microbial function genes

Following the completion of the experiment (a), an analysis of the intermediates of PFOA was conducted using LC-MS to elucidate the metabolic pathways and reaction mechanisms involved. Additionally, solid samples were collected for comprehensive characterization through techniques such as scanning electron microscope (SEM), X-ray diffraction (XRD), and X-ray photoelectron spectroscopy (XPS). These analyses aimed to investigate the changes in the secondary mineral species and surface morphology before and after the reaction. Furthermore, the PFOA-cultured microorganism samples in above Section (2) of this study were collected for metagenomics analysis. The samples were sequenced using Illumina PE150 (Illumina Inc.) instruments at Shanghai Meiji Biological Company. The microbial diversity, metabolic pathways, functional genes, and corresponding enzymes were investigated and analyzed using the Non-Redundant (NR) and Kyoto Encyclopedia of Genes and Genomes (KEGG) databases. The raw data (PRJNA908132) were stored at the National Center of Biotechnology Information (NCBI). The above analytical methods were provided as follows.

PFOA quantification was carried out using LC-MS instrumentation (LCMS-8040, SHIMADZU, Japan). To analyze the intermediates of PFOA, LC-MS analysis was performed using UPLC 1290 coupled with QTOF 6550 (Agilent, USA). The concentration of chloride ions was determined using selective electrodes specifically designed for chloride detection. Surface morphology and elemental mapping were examined using a SEM (SU8010, Hitachi, Japan). Elemental analysis was conducted using XPS (EscaLab 250Xi, Thermo Fisher Scientific, USA). The identification of secondary mineral species was accomplished through XRD analysis using a model Advance D8 instrument (Bruker, Germany).

(5) Microbe-Fe^0 interaction model for PFOA in vertical reaction column systems

The water flow parameters are shown in Table 5.3.

Table 5.3 The water flow parameters in the vertical reaction column system, which were obtained by the neural network prediction in the HYDRUS-1D.

Pedotransfer Function	Parameter
Input	
Sand (%)	100
Silt (%)	0
Clay (%)	0
Bulk Density ($g \cdot cm^{-3}$)	1.13
Output	
Qr (-)	0.0524
Qs (-)	0.5013
Alpha (cm^{-1})	0.0456
n (-)	3.0374
Ks ($cm \cdot h^{-1}$)	90.0*
l (-)	0.5

* The saturation permeability coefficient *Ks* was modified to 90.0 $cm \cdot h^{-1}$ based on the PFOA and Cl^- breakthrough curves; the predicted *Ks* was 45.2267 $cm \cdot h^{-1}$.

The parameters for PFOA and tracer are shown in Tables 5.4 and 5.5. The breakthrough curves of PFOA in microbe-Fe^0 interaction columns were characterized by Hydrus-1D. The two-site sorption coupled with the biological-chemical reaction model was developed to reveal the reaction behavior difference of PFOA. The governing equation is modified and defined as (Zhu et al., 2022):

$$\frac{\partial C}{\partial t} + f\frac{\rho}{\theta}K_d\frac{\partial C}{\partial t} + (1-f)\frac{\rho}{\theta}\frac{\partial S}{\partial t} = D\frac{\partial^2 C}{\partial z^2} - v\frac{\partial C}{\partial z} + \sum_k (k_{b_k} + \lambda_k)C, \quad (5-1)$$

$$\frac{\rho}{\theta}\frac{\partial S}{\partial t} = \alpha C. \quad (5-2)$$

where C represents the concentration of PFOA, t represents time, f represents the proportion of instantaneous retardation to total retardation, ρ represents the dry bulk density ($M \cdot L^{-3}$), θ represents water content, K_d represents the instantaneous equilibrium adsorption coefficient ($L^3 \cdot M^{-1}$), S represents the concentration of PFOA in the solid phase ($M \cdot M^{-1}$), D represents the hydrodynamic dispersion coefficient ($L^2 \cdot T^{-1}$), z represents the vertical spatial coordinate (L), and v represents the velocity of pore water ($L \cdot T^{-1}$), λ represents the rate of chemical reaction (T^{-1}), k_b represents the biodegradation rate [T^{-1}], α represents the first-order sorption coefficient (T^{-1}).

Table 5.4 Parameters for PFOA in different types of vertical reaction column systems.

Variable	Control	alone Fe^0	alone microbe	microbe-Fe^0
Disp (a/cm)	0.53	0.48	0.35	0.20
FRAC (f/–)	55%	48%	46%	40%
KD ($K_d/cm^3 \cdot g^{-1}$)	0.22	0.42	0.43	0.46
SNKL1 (λ or k_b/h^{-1})	—	0.15	0.18	0.22
ALPHA (α/h^{-1})	0.02	0.26	0.35	0.48
R^2	93.90%	96.19%	96.06%	97.07%

* f represents the proportion of instantaneous retardation to total retardation; first-order kinetic sorption fraction $f' = 1-f$.

Table 5.5 Parameters for Cl^- in different types of vertical reaction column systems.

Variable	Control	alone Fe^0	alone microbe	microbe-Fe^0
Disp (a/cm)	0.53	0.48	0.35	0.20
FRAC (f/−)	60%	17%	10%	4%
KD ($K_d/cm^3 \cdot g^{-1}$)	0.14	0.12	0.26	0.35
ALPHA (α/h^{-1})	0.11	0.10	0.14	0.21
R^2	98.96%	99.27%	99.37%	99.85%

5.2.3 Results and discussion

(1) The PFOA removal kinetics in the batch microbe-Fe^0 interaction system

The results revealed that the microbe-Fe^0 interaction system was the most effective in PFOA removal, with a 48% PFOA removal achieved in 8 days (Figure 5.4). This removal rate was distinctly higher than that in the alone Fe^0 (22%) and alone microbe (31%) systems. The removal performance of the microbe-Fe^0 interaction system increased by 55%–118%. Moreover, the removal of PFOA met the first-order kinetic model (Wang et al., 2020). Compared to the alone Fe^0(0.0076 day^{-1}) and alone microbe (0.0172 day^{-1}) system, the reaction rate k (0.0426 day^{-1}) of PFOA in the microbe-Fe^0 interaction system increased by 1.5 to 4.6 folds, respectively. Furthermore, more F^-(0.355 $mg \cdot L^{-1}$) was produced in the microbe-Fe^0 interaction system than that (0.190 $mg \cdot L^{-1}$) in the alone microbe system (Figure 5.5). It was demonstrated that microbe-Fe^0 interaction can achieve greater reaction rate k and removal performance for PFOA. The detailed mechanism will be explained in the following Section (5) of this study.

(2) PFOA transport and reaction behavior in different types of vertical reaction column systems

Multi-source data (Cl^- and PFOA) were combined to calibrate and adjust the model parameters. The dispersion a (Tables 5.4 and 5.5) and hydraulic conductivity

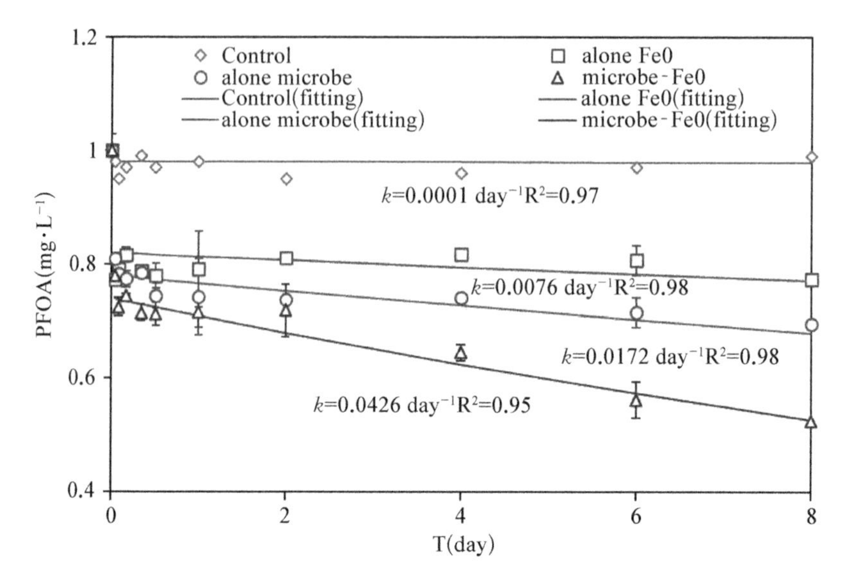

Figure 5.4 PFOA-removal kinetics in control, alone Fe^0, alone microbe, and microbe-Fe^0 batch reaction systems. No amendment was added to the control.

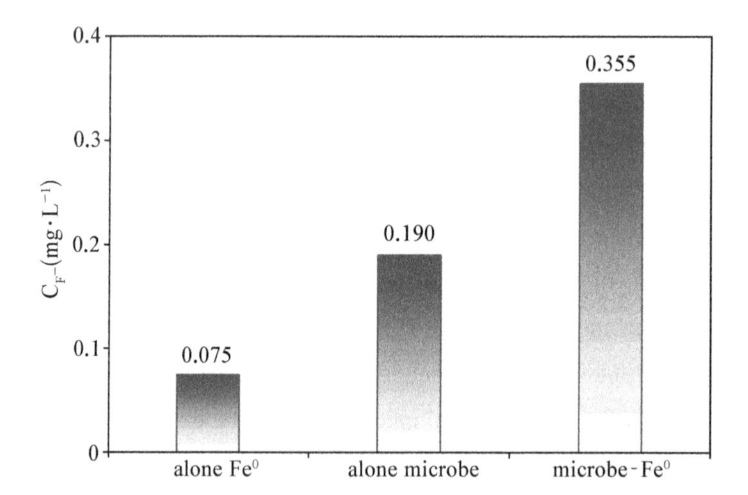

Figure 5.5 F^- concentration in alone Fe^0, alone microbe, and microbe-Fe^0 interaction batch systems.

*K*s (Table 5.3) were validated based on the breakthrough curves of PFOA (Figure 5.6 (a)) and Cl^-(Figure 5.6 (b)). The results indicated that the distribution coefficient K_d(0.14 $cm^3 \cdot g^{-1}$ to 0.35 $cm^3 \cdot g^{-1}$) and first-order kinetic sorption coefficient α (0.11 h^{-1} to 0.21 h^{-1}) for Cl^- increased significantly in the microbe-Fe^0 interaction columns (Table 5.5).

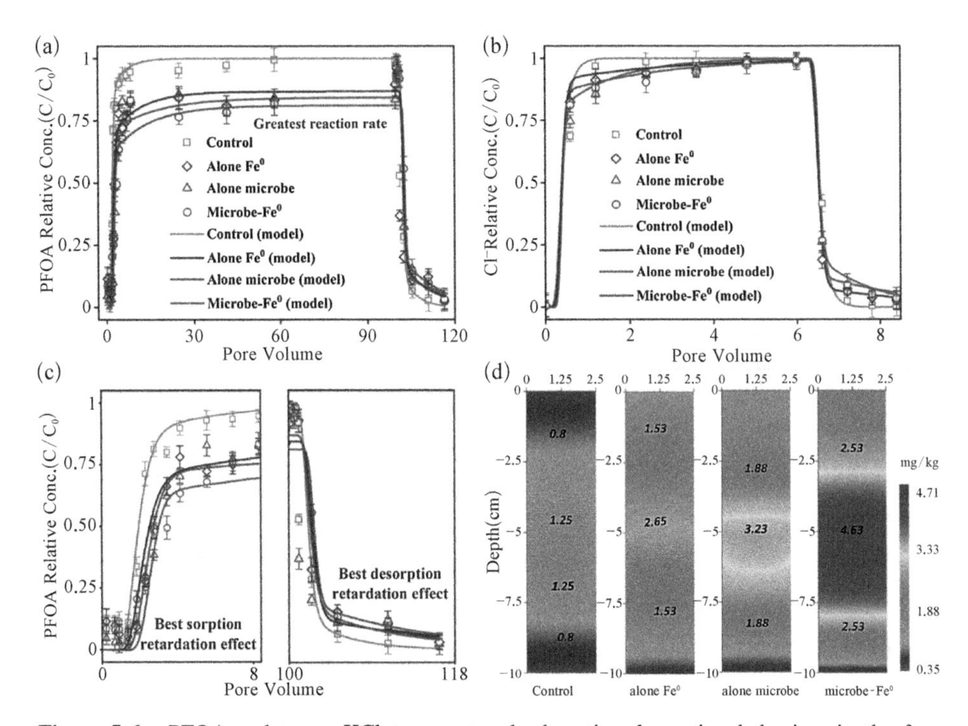

Figure 5.6 PFOA and tracer KCl transport and adsorption-desorption behaviors in the four types of vertical reaction column systems (a)-(b); sorption and desorption curves change of PFOA (c) and 2D-spatial distribution (d) of PFOA in solid media with residual water.

Figure 5.6 (a) showed that 76% of PFOA breakthrough in the microbe-Fe^0 interaction columns, which was lower than the breakthrough percentages observed in the alone microbe (79%), alone Fe^0(82%), and Control (100%). Additionally, a distinct change was found in the sorption-desorption curves in 0 – 10 PVs and 100 PVs – 120 PVs (Figure 5.6 (c)), suggesting the best removal performance, sorption-desorption retardation effect for PFOA in the microbe-Fe^0 interaction columns. Figure 5.6 (d) shows the actual solid phase distribution in Control, alone Fe^0, alone microbe, and microbe-Fe^0 interaction columns. PFOA was predominantly detected at the material filling site. The adsorption capacity (4.80 $mg \cdot kg^{-1}$) in the microbe-Fe^0 interaction column was significantly greater than that in the Control (1.25 $mg \cdot kg^{-1}$), alone Fe^0(2.60 $mg \cdot kg^{-1}$), and alone microbe (3.30 $mg \cdot kg^{-1}$) systems.

The difference of model parameters (Table 5.4) further revealed the potential mechanism: compared to alone Fe^0 column, the dispersion a decreased (0.48 cm to 0.20 cm), while K_d(0.42 $cm^3 \cdot g^{-1}$ to 0.46 $cm^3 \cdot g^{-1}$) and λ (0.15 h^{-1} to 0.22 h^{-1}) increased significantly in the microbe-Fe^0 interaction column. In addition, the coefficient α (0.26 h^{-1} to 0.48 h^{-1}) and fraction f' (52% to 60%) of first-order kinetic sorption also showed an increase. Taken together, the microbe-Fe^0 interaction columns exhibited the best sorption and desorption retardation effects, as well as the highest removal efficiency for PFOA, when compared to the other two types of vertical reaction columns.

(3) Secondary minerals change at CZ gel beads interface in different reaction systems

Before the reaction (Figures 5.7 (a)-(b)), the surface of CZ appeared smoothly and partially porous, and SEM element mapping indicated the uniform distribution of Fe and O and the presence of some iron oxides (Figure 5.7 (c)-(f)).

After the reaction, needle-like and rod-like secondary minerals were observed (Figure 5.7 (h)), consistent with the morphological characteristics of akaganeite (β-FeOOH) and XRD results (Figure 5.8 (g)). Furthermore, SEM images showed severe corrosion of the smooth surface of CZ with the formation of flocculent iron oxides (Figure 5.7 (j)). Figures 5.8 (a)-(f) showed XPS analysis results of CZ gel beads. The peak of Fe $2p_{3/2}$ at 710.1 eV and Fe $2p_{1/2}$ at 724.0 eV is Fe^{2+}. The peak of Fe $2p_{3/2}$ at 712.5 eV and Fe $2p_{1/2}$ at 727.0 eV is Fe^{3+}(Rajajayavel and Ghoshal, 2015; Zazpe et al., 2023). The results showed that Fe^{3+} (30.77%) and Fe^{2+} (69.23%) oxides existed before the CZ reaction (Figure 5.8 (a)). In addition, iron hydroxide (Fe-OH) and iron oxide (Fe-O) were found in the O 1s range (Figure 5.8 (b)) with peaks at 529.6 and 531.2 eV representing Fe-OH (86.42%) and Fe-O species (13.58%) (Figure 5.8 (b)) (Kim et al., 2011a). XPS analysis revealed the presence of Fe^{3+}(38.79%) and Fe^{2+}(61.21%) oxides in the CZ from the alone Fe^0 system after the reaction (Figure 5.8 (c)).

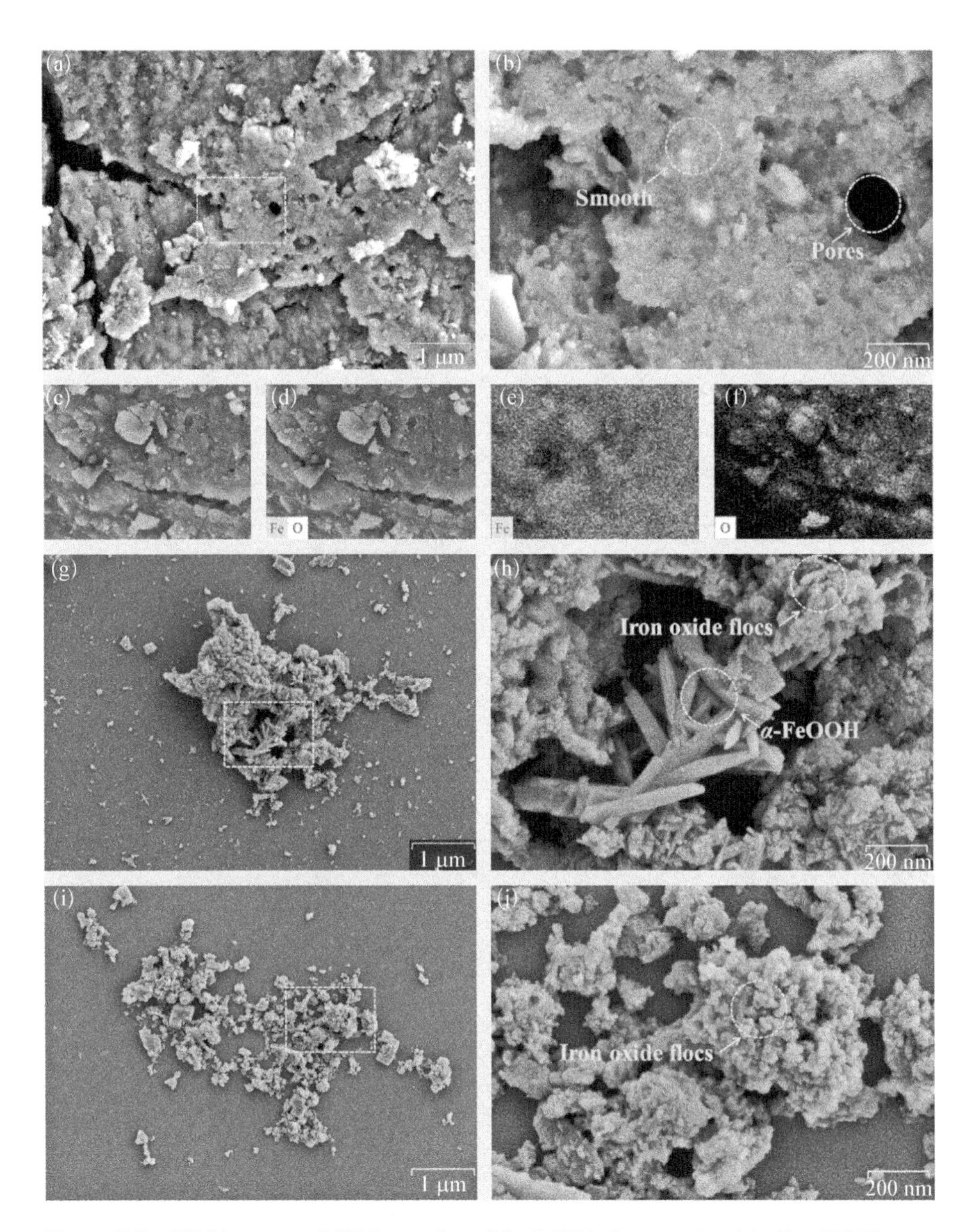

Figure 5.7 SEM images and SEM-mapping of fresh CZ before reaction (a)-(f); SEM images of CZ from alone Fe^0 system after reaction (g)-(h); SEM images of CZ from the microbe-Fe^0 interaction system after reaction (i)-(j).

Additionally, in the microbe-Fe^0 system, XPS results confirmed the presence of Fe^{3+}(52.49%) and Fe^{2+}(47.51%) oxides in CZ after the reaction, indicating the residual of Fe^{2+} oxides (Fe_3O_4 confirmed by XRD) was less than that in alone Fe^0 system after the reaction (Figure 5.8 (e)). Furthermore, XPS analysis displayed three peaks at 532.9 eV, 531.4 eV, and 530.1 eV, corresponding to C = O (34.04%), Fe-OH (54.76%), and Fe-O (11.20%) at the CZ interface (Figure 5.8 (d)) (Kim et al., 2011a), respectively. In the microbe-Fe^0 interaction system, the XPS analysis assigned the three peaks to C=O (51.79%), Fe-OH (39.72%), and Fe-O (8.50%) at the CZ interface (Figure 5.8 (f)). The presence of C=O presumably comes from the byproduct $F(CF_2)_6FC=O$ of PFOA degradation, as confirmed by the results of LC-MS (Figure 5.11 (c)).

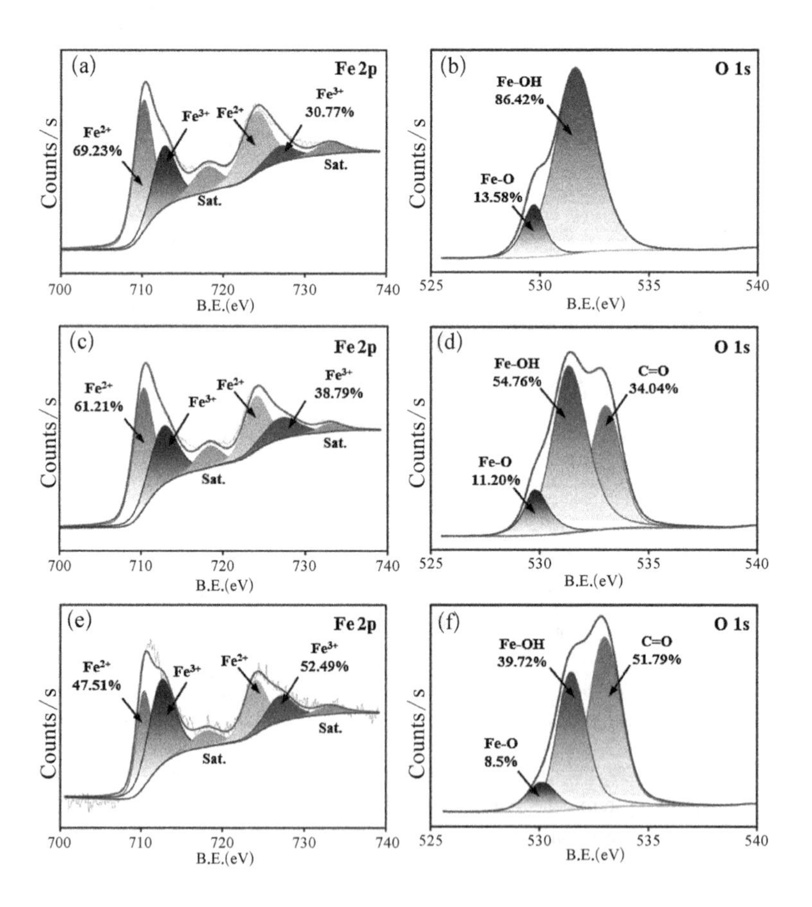

(g)

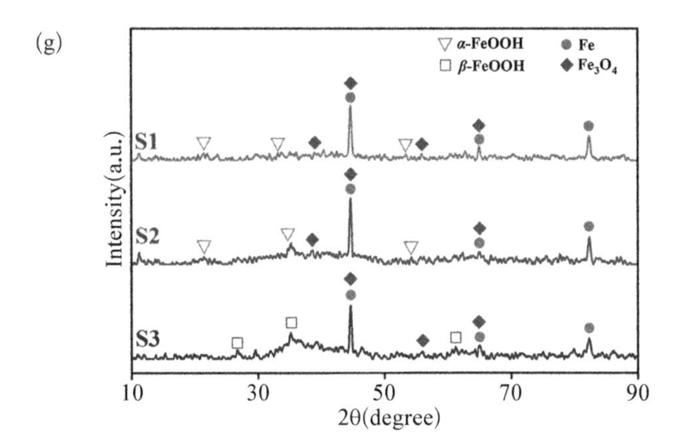

Figure 5.8 XPS spectra: Fe 2p and O 1s region of CZ before reaction (a)-(b); Fe 2p and O 1s region of CZ from alone Fe^0 system after reaction (c)-(d); Fe 2p and O 1s region of CZ from microbe-Fe^0 interaction system after reaction (e)-(f). Sat. represents the satellite peak. XRD patterns (g); S1 represented the before reaction; S2 and S3 represented the CZ samples from the alone Fe^0 and microbe-Fe^0 interaction systems after the reaction, respectively.

XRD analysis results (Figure 5.8 (g)) showed that the CZ amendment mainly consists of Fe^0(JCPDS #6-696), Fe_3O_4(JCPDS #26-1136) and α-FeOOH (JCPDS #74-2195). This observation was consistent with XPS results for the presence of Fe-O (13.58%) and Fe-O-H (86.42%) (Figure 5.8 (a)), which could be attributed to partial oxidation of the CZ amendment during preparation. Fe^{2+} underwent oxidation to Fe^{3+}(Figures 5.8 (c) and 5.8 (e)) during the process of PFOA removal. The main minerals detected in the alone Fe^0 system were Fe_3O_4 (JCPDS#26-1136) and α-FeOOH (JCPDS#74-2195) (Figure 5.8 (g)). The analysis of XPS confirmed that the percentage of Fe^{3+} changed from 30.77% to 38.79% (Figures 5.8 (a) and 5.8 (c)) and Fe-OH and Fe-O changed to 54.76% and 11.20%, respectively (Figure 5.8 (d)). Whereas, the main mineral was β-FeOOH (JCPDS#1-662) with less residual Fe_3O_4 in the microbe-Fe^0 interaction system (Figure 5.8 (g)). Interaction between the microbe and Fe^0 influenced the fate of Fe^{2+} and thus favored the formation of β-FeOOH. Further details mechanism was shown in the following Section (5) of this study.

(4) Functional genes for PFOA degradation and corresponding metabolic pathways and enzymes

The microbial community at the phylum level consisted primarily of

Proteobacteria, *Euryarchaeota*, *Actinobacteria*, *Chloroflexi*, *Firmicutes*, *Bacteroidetes*, and *Armatimonadetes* (Figure 5.9 (a)). Specifically, the abundance of *Proteobacteria* (97.11%) increased by 41.12% under PFOA conditions than Control (68.81%) (Figure 5.9 (a)).

This increase in *Proteobacteria* abundance under PFOA conditions aligned with findings from Sun et al. who also observed that PFOA exposure can lead to a reduction in bacterial community diversity and an increase in *Proteobacteria* abundance (Sun et al., 2016). Under PFOA exposure, the community at the genus level was mainly (99.26%) composed of *Delftia*, *Achromobacter*, *Methanosarcina*, *Rhizobiales*, *Dehalococcoidia*, *Thiobacillus*, *Actinobacteria*, *Proteobacteria*, *Pseudomonas*, *Chloroflexi*, *Methanothrix*, and *Armatimonadetes*. Among these, *Delftia*, *Achromobacter*, *Pseudomonas*, and *Proteobacteria* were the dominant bacterial groups. Notably, the abundance of the genus *Pseudomonas* (1.63%) increased significantly by 472.6% under the PFOA condition compared to the Control (0.29%), with its relative proportion reaching 1.63% (Figure 5.9 (b)). Genus *Pseudomonas*, which belongs to the phylum *Proteobacteria*, has been reported to favor the transformation of fluoromodulated alcohols (FTOH) into less toxic short-chain PFCAs by the stepwise removal of -CF_2- moiety (Kim et al., 2012). Moreover, *Pseudomonas*, a key genus of iron-redox cycling bacteria (IORB), is capable of accelerating the generation of Fe^{2+} through the oxidation of Fe^0 or serving as an electron donor to the reduction of Fe^{3+} in the periplasm (Xu et al., 2020). The results above highlighted the presence of phylum functional microorganism *Proteobacteria* and genus *Pseudomonas*, which are associated with defluoridation and Fe^0transformation, in the PFOA domestication samples.

Compared to the Control, the results of KEGG level functional hierarchy showed that the number of metabolic genes in *Metabolism*, *Environmental information processing*, *Organismal systems*, and *Cellular processes* upregulated after PFOA domestication (Figure 5.9 (c)). The key pathways to focus on are the Xenobiotics biodegradation and metabolism pathway in the secondary functional hierarchy of the *Metabolism* pathway, in which the number of metabolic genes in the Xenobiotics biodegradation and metabolism (1.46%) pathway increased by

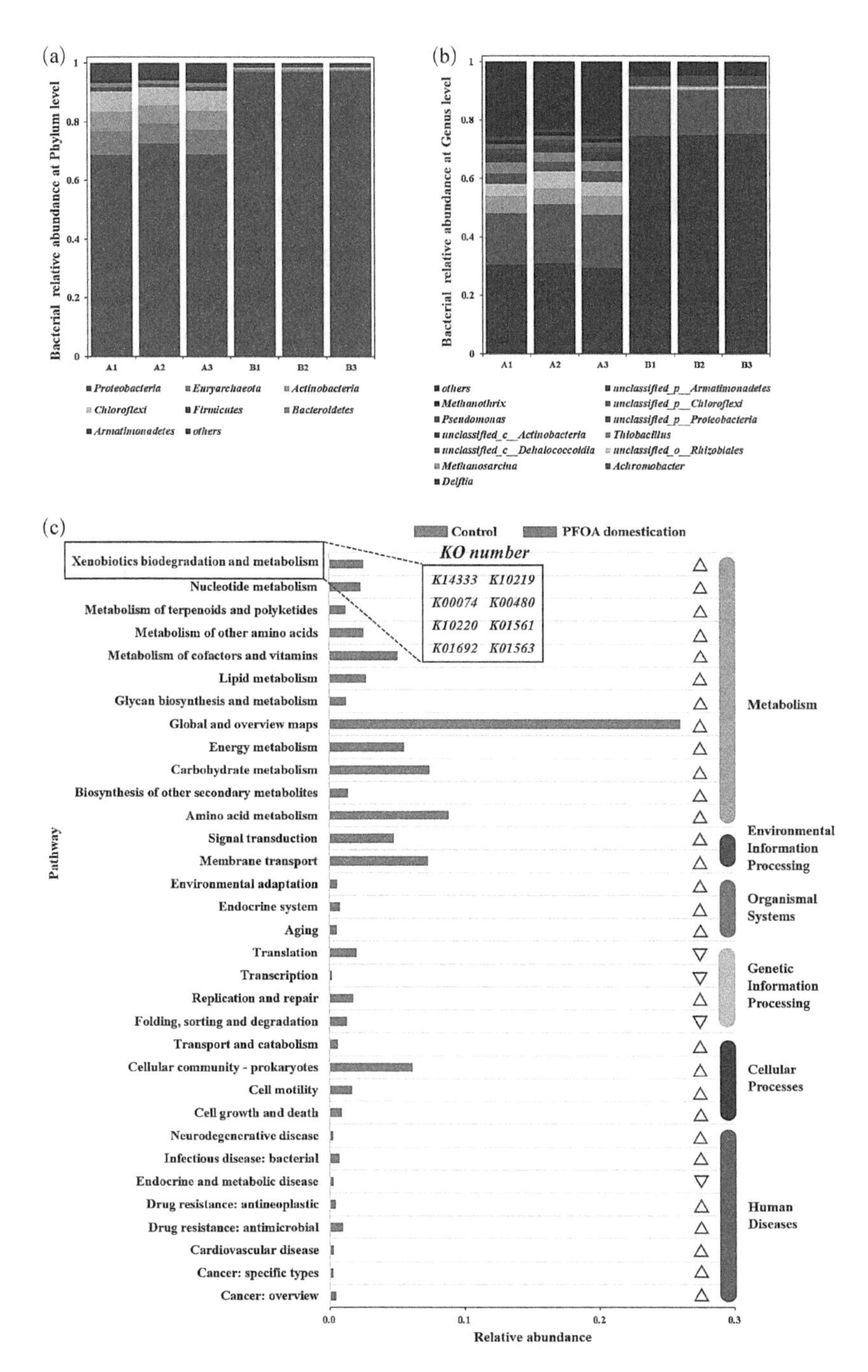

Figure 5.9 Diversity and abundance of bacteria at the phylum (a) and the genus (b) levels; differences in potential functions between the Control and PFOA domesticated microorganism (c). Pathways with red △ mean the abundance is increased; the blue ▽ represents the opposite condition.

32.7% compared to that of the Control (1.10%). This noteworthy upregulation (Figure 5.10) suggested the effective activation of microbial pathways for the biodegradation of xenobiotic substances after PFOA domestication. These substances, including benzyl esters, fluorobenzyl esters, and polycyclic aromatic hydrocarbons, represent ubiquitous organic pollutants of concern in the environment. Remarkably, some of these compounds share similar chemical structures and functional groups with the PFOA, indicating the possibility of analogous metabolic pathways and specific enzymes involved in PFOA degradation (Figure 5. 10 (a)). Furthermore, a study by Xiao et al. (Chakraborty et al., 2021) demonstrated that the primary pathway employed by microorganisms for the decomposition and mineralization of diverse xenobiotic substances is the benzoate degradation pathway (ko00362). Our KEGG annotation results (Table 5.6) corroborate these findings, showing a slight downregulation of the functional gene K14333 (*DHBD*), responsible for encoding decarboxylase, in the PFOA domestication group. Conversely, the functional genes K10220 (*ligJ*) for encoding hydratase, K00074 (*hbd*) for encoding alcohol dehydrogenase, and K10219 (*ligC*) for encoding aldehyde dehydrogenase were significantly upregulated in the same group. Differential analysis based on KEGG annotation results revealed distinct variations in the ko00362 pathway (Figure 5.10 (a)). During the enzymatic reactions involved in PFOA degradation decarboxylases facilitated the removal of carboxyl groups, followed by hydration reactions mediated by hydratases, resulting in the formation of decarboxyl alcohols. Subsequently, alcohol dehydrogenases catalyzed the oxidation of decarboxyl alcohols into aldehydes (or ketones), which were further transformed into carboxylic acids by aldehyde dehydrogenases. Such enzymatic decarboxylation represents one of the pivotal biochemical processes within living organisms (Kim et al., 2021). Moreover, Tiedt et al. (Tiedt et al., 2017) highlighted the involvement of the functional gene K01692, responsible for encoding the enoyl-CoA hydratase enzyme, in the enzymatic defluorination of 2-F-benzoate. Encouragingly, our KEGG annotation results (Table 5.6) confirmed the presence of this functional gene K01692 in the microbial samples analyzed in this study.

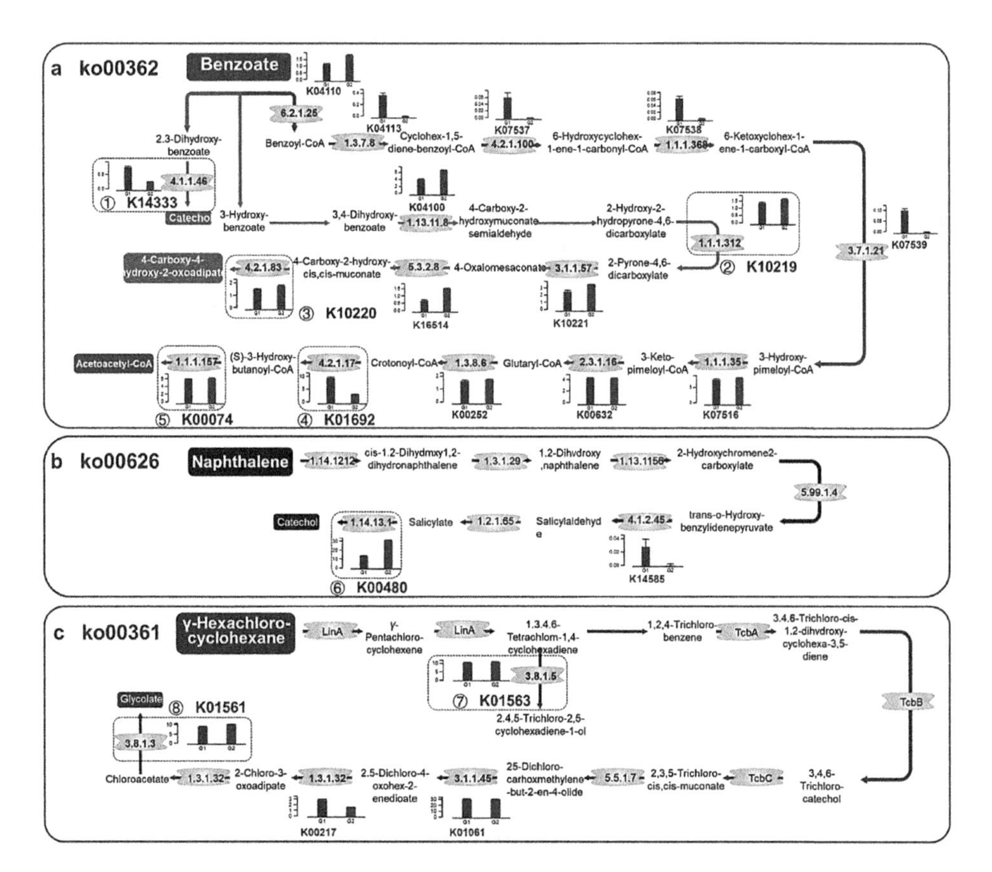

Figure 5.10 Functional genes/enzymes involved in the benzoate degradation pathway (ko00362) (Kim et al., 2021; Tiedt et al., 2017) (a), naphthalene degradation pathway (ko00626) (Van Beilen et al., 2006) (b), chlorocyclohexane and chlorobenzene degradation pathway (ko00361) (Xiao et al., 2023) (c) were obtained based on KEGG. Similar pathways (functional genes and specific enzymes) are also involved in PFOA metabolism; the functional genes encoding key enzymes are highlighted in red.

Table 5.6 Abundance variations of functional gene fragments related to PFOA degradation, metabolic pathways and enzymes.

Pathway ID	KEGG Orthology (KO) number	Enzyme type	A1-Control	A2-Control	A3-Control	B1-PFOA domestication	B2-PFOA domestication	B3-PFOA domestication
ko00362	K14333	2,3-dihydroxybenzoate decarboxylase	1 475	1 940	1 872	1 154	1 216	1 016
	K10220	4-oxalmesaconate hydratase	4 270	3 606	3 172	***6 726****	***6 532****	***6 158****
	K00074	3-hydroxybutyryl-CoA dehydrogenase	18 708	16 492	14 564	***24 828****	***25 038****	***21 646****
	K10219	2-hydroxy-4-carboxymuconate semialdehyde hemiacetal dehydrogenase	4 078	3 832	3 148	***6 386****	***6 476****	***5 418****
	K01692	enoyl-CoA hydratase	25 312	21 344	19 954	11 936	12 272	10 750
ko00626	K00480	salicylate hydroxylase	4 250	5 288	4 696	***13 164****	***11 426****	***12 948****
ko00361	K01563	haloalkane dehalogenase	4 694	4 062	3 456	***6 926****	***7 346****	***6 370****
	K01561	haloacetate dehalogenase	3 852	3 586	2 918	***6 674****	***6 668****	***5 948****

* *indicates that the number of coding gene fragments is up-regulated.*

In addition to these findings, the study by Van et al. (Van Beilen et al., 2006) provided compelling evidence regarding the role of the hydroxylation functional gene K00480, responsible for encoding salicylate hydroxylase, in the degradation of trichloroethylene through terminal hydroxylation of alkyl chains. Notably, our results in Table 5.6 demonstrated a significant increase in the relative abundance of the functional gene K00480 in the microbial samples after PFOA domestication. The differential analysis further suggested the occurrence of the K00480 gene fragment within the naphthalene degradation pathway (ko00626) (Figure 5.10 (b)). Moreover, the groundbreaking study by Goldman et al. (Goldman, 1965) uncovered the efficacy of fluorinated acyl-CoA hydrolase, isolated from *Pseudomonas*, in catalyzing the hydrolytic defluorination of fluorinated compounds. Remarkably, the hydrolysis mediated by dehalogenase enzymes facilitates the cleavage of the C-F bond in various fluorinated aromatic compounds, such as fluorobenzene (Carvalho et al., 2005) and fluorobenzoate (Chae et al., 2008). These hydrolytic reactions are governed by the functional genes K01563 (*dhaA*) and K01561 (*dehH*), encoding haloalkane dehalogenase and haloacetate dehalogenase, respectively (Xiao et al., 2023). Importantly, our results in Table 5.6 unveil a significant upregulation in the abundance of the functional genes K01563 (*dhaA*) and K01561 (*dehH*) following PFOA domestication. Notably, the differential analysis suggested the incorporation of the K01563 (*dhaA*) and K01561 (*dehH*) gene fragments within the chlorocyclohexane and chlorobenzene degradation pathway (ko00361) (Figure 5.10 (c)).

(5) PFOA reaction pathway and secondary mineral transformation in microbe-Fe^0 interaction system

Based on the LC-MS results and previous research, the possible degradation pathways of PFOA are proposed to encompass a series of processes (Figure 5.11): A: Activation (Liu et al., 2021), B: Deprotonation (Liu et al., 2021), C: Decarboxylation (Liu et al., 2021), D: Hydration (Liu et al., 2021), E: F-elimination (Liu et al., 2021), F: C-C bond scission (Ding et al., 2021), G: Decarboxylase (Xiao et al., 2023), H: Dehalogenase (Xiao et al., 2023), I: Hydratase (Xiao et al., 2023)/Hydroxylase (Van Beilen et al., 2006), J: Alcohol dehydrogenase (Xiao et al., 2023), K: Aldehyde dehydrogenase (Kim et al., 2021; Wackett, 2022), L: Enoyl-CoA hydratase (Tiedt et al., 2017), and M: HF elimination (Liu et al., 2021). The degradation pathway in the alone Fe^0 system was shown in Figure 5.11 (a). The PFOA underwent activation (A) to generate F-$(CF_2)_7$-COOH・, accompanied by simultaneous deprotonation (B) to form F-$(CF_2)_7$-COO^-. Subsequently, decarboxylation (C) took place, leading to the production of F-$(CF_2)_7$・ and CO_2. The detection of the intermediate F-$(CF_2)_7$・ provided evidence for the occurrence of the decarboxylation reaction. Ensuing Hydration (D) and F^- elimination (E) reactions were followed by the formation of the intermediate F-$(CF_2)_6$-CF-OH・ (Liu et al., 2021). Additionally, the detection of CF_3COOH, in accordance with findings of Ding et al. (Ding et al., 2021), suggested that PFOA can undergo degradation through F^- elimination (E) and spontaneous α-C-C bond cleavage (F), yielding short-chain CF_3COOH.

Figure 5.11 (b) represented the degradation pathway in the alone microbe system. Initially, the microbial system attacked the -COOH group, catalyzed by carboxylic acid dehydrogenase, leading to the formation of F-$(CF_2)_7$・. The detection of the intermediate F-$(CF_2)_7$・ confirmed the occurrence of the enzyme-mediated decarboxylation (G) reaction. Subsequently, dehalogenase (H) (Xiao et al., 2023) process occurred, resulting in the production of F-$(CF_2)_5$-$CF=CF_2$. Under the catalysis of hydratase (Kim et al., 2021)/hydroxylase (I) (Van Beilen et al., 2006), the corresponding 6-perfluorohexyl-1-ol (F-$(CF_2)_6$-CH_2-OH) was generated. It then underwent a reaction with alcohol dehydrogenase (J) to form the corresponding 6-perfluorohexyl-1-aldehyde (F-$(CF_2)_5$-CF_2-CH=O), which was further oxidized by aldehyde dehydrogenase (K) to produce 6-perfluorohexyl-1-carboxylic acid (F-$(CF_2)_5$-CF_2-COOH) (Kim et al., 2021; Wackett, 2022). Subsequent cyclic reactions (G, H, I, J, K) occured, and the detection of the intermediate product CF_3COOH confirmed this degradation pathway.

Figure 5.11 c illustrated the degradation pathway of PFOA in the microbe-Fe^0 interaction system. LC-MS results revealed the detection of several intermediates, including F-$(CF_2)_7$・, F-$(CF_2)_6$-COOH, short-chain CF_3COOH, suggesting a novel degradation pathway under the synergistic action of microbes and Fe^0. In this pathway III, PFOA underwent decarboxylation (C) to produce F-$(CF_2)_7$ ·, which can be catalyzed by enoyl-CoA hydratase (L) to form F-$(CF_2)_7$-OH (Tiedt et al., 2017). Subsequently, HF elimination (M) took place to yield the product F-$(CF_2)_6$FC=O.

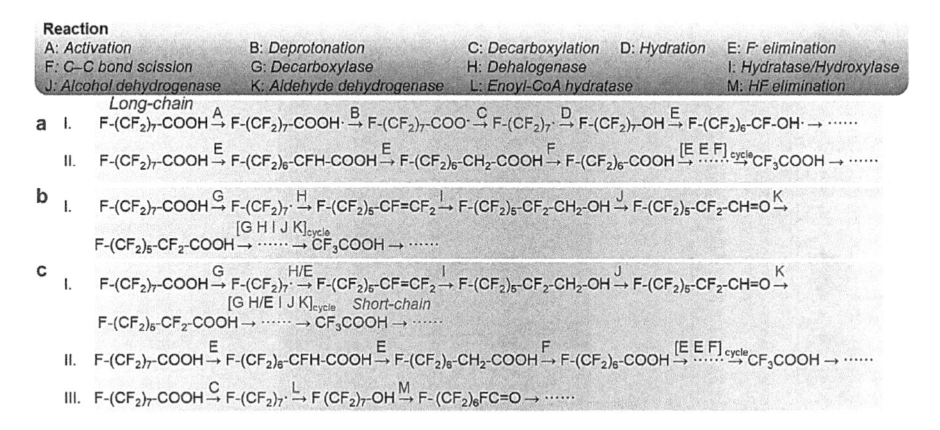

Figure 5.11 PFOA reaction pathways in alone Fe^0(Ding et al., 2021; Liu et al., 2021) (a), alone microbe (Kim et al., 2021; Wackett, 2022) (b), and microbe-Fe^0 interaction (Tiedt et al., 2017) (c) systems. The red marks indicate that the intermediates were detected by LC-MS in this study. Red characters represent general chemical processes, while green characters indicate the involvement of enzymes in biological processes in the reaction A-M.

Based on the results of Figures 5.7 – 5.11 and previous reports (Ding et al., 2021; Fan et al., 2021; Kim et al., 2021; Liu et al., 2021; Tiedt et al., 2017; Van Beilen et al., 2006; Wackett, 2022; Xiao et al., 2023; Xu et al., 2020), the proposed transformation paths of iron oxides and the reaction mechanism of PFOA in alone Fe^0 and microbe-Fe^0 interaction systems were presented in Figure 5.12. In the alone Fe^0 system, during the removal process of PFOA, Fe^{2+} undergoes oxidation to Fe^{3+}. Under the catalytic action of ZVI, PFOA undergoes a series of reduction-defluorination and carbon chain cleavage reactions, exhibiting superb green cycling strategies involving key steps such as *Activation*, *Deprotonation*, *Decarboxylation*, *Hydration*, *F^- elimination* (Liu et al., 2021) and [*F^- elimination*; *C C bond scission* (Ding et al., 2021)]$_{cycle}$ process. These complex reaction chains eventually result in the generation of F-$(CF_2)_6$-CF-OH · and short-chain CF_3COOH.

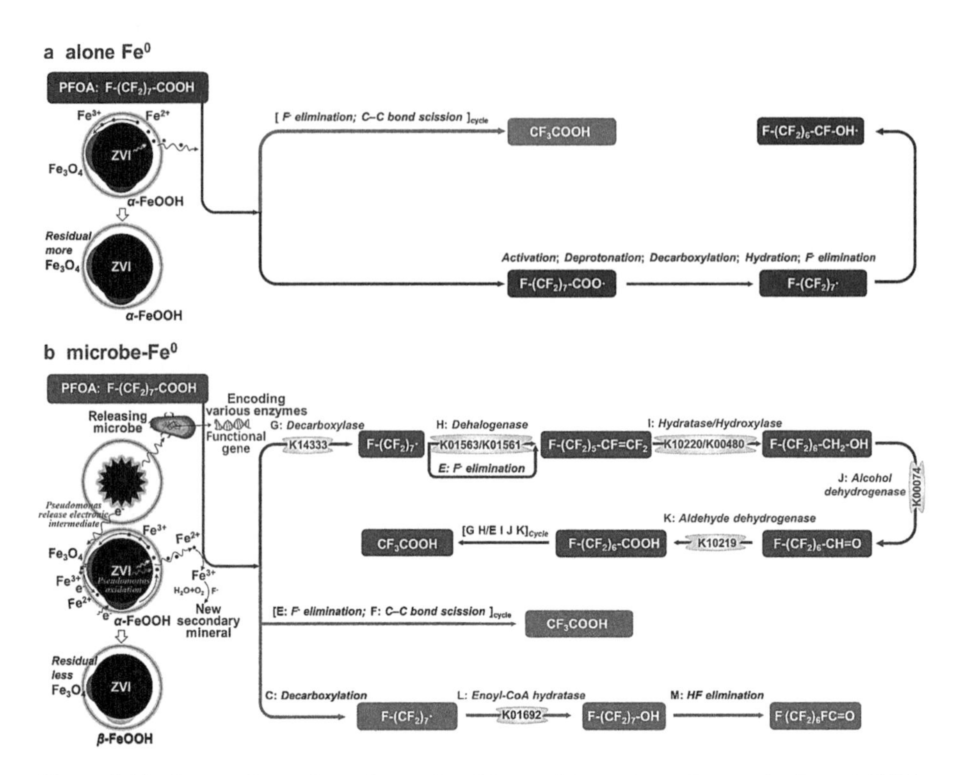

Figure 5.12 Proposed transformation paths of iron oxides and reaction mechanism of PFOA in alone Fe^0(Ding et al., 2021; Fan et al., 2021; Liu et al., 2021) (a), and microbe-Fe^0 interaction (Kim et al., 2021; Tiedt et al., 2017; Van Beilen et al., 2006; Xiao et al., 2023; Xu et al., 2020) (b) systems based on our research and previous reports.

However, in the microbe-Fe^0 interaction system, *Pseudomonas*, a key genus of functional bacteria, plays a crucial role in the removal of PFOA. *Pseudomonas* accelerates the oxidation of $Fe^0$to generate Fe^{2+} and serves as an electron donor to the reduction of Fe^{3+}(Xu et al., 2020). In this process, the secondary mineral magnetite (Fe_3O_4) at the interface is removed (Figure 5.8 (g)) probably under the effect of *Pseudomonas*. In the presence of F^-(Figure 5.5) or Cl^-, β-FeOOH can be produced (Chitrakar et al., 2006) or α-FeOOH can be transformed into β-FeOOH (Fan et al., 2021). PFOA undergoes a series of enzymatic defluorination, reduction-defluorination, and carbon chain cleavage reactions through the following cycles: [*Decarboxylase* (K14333) (Kim et al., 2021), *Dehalogenase* (K01563/K01561) (Xiao et al., 2023)/*F^- elimination* (Xiao et al., 2023),

Hydratase (K10220) (Kim et al., 2021)/*Hydroxylase* (K00480) (Van Beilen et al., 2006), *Alcohol dehydrogenase* (K00074) (Kim et al., 2021), *Aldehyde dehydrogenase* (K10219) (Kim et al., 2021; Wackett, 2022)]$_{cycle}$, [*F^- elimination* (Xiao et al., 2023), *C C bond scission* (Ding et al., 2021)]$_{cycle}$ and *Decarboxylation* (Liu et al., 2021), *Enoyl-CoA hydratase* (K01692) (Tiedt et al., 2017), *HF elimination* (Liu et al., 2021) processes. These enzymatic superb green cycling strategies result in the formation of short-chain CF_3COOH and F-$(CF_2)_6$FC=O.

(6) Conclusion of this case study

The integration of microbe-Fe^0 interactions, enzymatic pathways mediated by key genes, and mineral transformations present a superb green cycling strategy for efficient PFOA elimination. In this study, the high-performance and green low-carbon CZ and CB amendments were developed to construct the microbe-Fe^0 high-rate interaction systems. The PFOA removal rate in the microbe-Fe^0 interaction batch system increased by 1.5 folds to 4.6 folds compared to that in alone ZVI or microbe systems. The presence of *Pseudomonas* accelerated the transformation of Fe^0 and Fe^{3+}, significantly influencing the PFOA transport and reaction behaviors. The multi-process coupling model demonstrated that microbe-Fe^0 interaction improved the retardation effect for the PFOA in vertical columns, in which the microbe-Fe^0 interaction decreased the dispersivity a; improved the distribution coefficient K_d, the reaction rate λ, and the fraction f' of first-order kinetic sorption for PFOA. Moreover, the LC-MS results revealed that microbe-Fe^0 interaction increased the diversity of PFOA reaction pathways. Furthermore, three related key metabolic pathways, eight key functional genes and their corresponding enzymes for the PFOA degradation were identified and proposed through metagenomics. These investigations greatly enhance our understanding of exceptional green cycling strategies for microbe-Fe^0 "neural network-type" interaction, which rely on the involvement of eight key genes encoding enzymes and mineral transformation processes, collectively contributing to eco-friendly and cost-effective methods for addressing the PFOA contamination.

5.3 Case study for TBBPA elimination by microbe-aged Fe^0 interaction

5.3.1 Research background

Tetrabromobisphenol A (TBBPA) is one of the most widely applied brominated flame retardants (BFRs) with the highest output in the world, and its usage accounts for 60% of the BFRs consumption (Law et al., 2006). TBBPA is primarily applied as an additive BFRs for textiles, plastics, etc (Liu et al., 2016). Due to such a widespread application, TBBPA is widely detected in sediment (Wang et al., 2021), soil (Jeon et al., 2021), water (Li et al., 2019), and air (Ma et al., 2021). The maximum concentration of the criterion of TBBPA is $1.\tilde{0}$ mg · kg^{-1} b.w. by UK Committee (Yu et al., 2019), while in China, the criterion maximum concentration of TBBPA in aquatic environments is 0.1475 mg·L^{-1} (Feiteiro et al., 2021). TBBPA would accumulate in aquatic organisms, birds, and mammals through the food chain (Gu et al., 2020), resulting in neurotoxic, nephrotoxic, and reproductive toxic effects (Yu et al., 2021). Therefore, it is urgent to develop effective methods and technologies to eliminate TBBPA and thus ensure environmental health.

In recent years, biological degradation (Peng et al., 2017), chemical reduction (Liu et al., 2019), physical adsorption (Cui et al., 2019), and advanced oxidation technology (Xiang et al., 2021a) have been proposed for TBBPA and other contaminants Congo Red (CR) dye (Kamran et al., 2019) and lithium (Kamran and Park, 2020) removal. Amongst them, physical adsorption and chemical reduction of nanoscale zero-valent iron (NZVI) are the most promising technologies to remove TBBPA, due to the vigorous surface activity and large specific surface area of NZVI (Liu et al., 2019). However, the NZVI powder is easy to aggregate, which reduces the surface area available for reaction, resulting in a reduction in the TBBPA removal efficiency. Researchers have proposed the preparation of composite materials by loading NZVI onto graphene oxide (GO) (Wang et al., 2015), biochar (Zhang et al., 2021), and multiwalled carbon nanotubes (CNTs) (Xiang et al., 2020) to reduce aggregation and improve the performance of TBBPA removal. In particular, GO has a large surface area with oxygen-containing surface functional groups, which can be easily synthesised using simple and eco-friendly/economic chemical procedures (Kamran et al., 2020). Compared to the NZVI powder, the NZVI-loaded GO (NZVI/GO) has shown an effective improvement in the reaction rate of pollutants (Chen et al., 2016). Although the performance of NZVI/GO improved significantly, a surface iron oxide/hydroxide passivation layer formed after the long-term application of NZVI hinders electron transfer between the inner NZVI and surface contaminants.

To overcome this limitation, iron-reducing bacteria were introduced into the system to decompose the surface iron oxide/hydroxide passivation layer associated with aged NZVI/GO so that the reaction sites could be reactivated and the performance of the amendments could be improved (Xiang et al., 2020). Iron-reducing bacteria utilized Fe(III) oxides as electron acceptors to produce adsorbent Fe(II) oxides or remove NZVI surface passivates by reducing Fe(III) oxides to Fe(II) species (Li et al., 2019). Several studies have shown that the *Schewanella* iron-reducing bacteria can reactivate aged NZVI and significantly improve the removal rate of trichloroethene (TCE) (Yang et al., 2017), uranium (U(VI)) (Xiang et al., 2020), *p*-nitro-chlorobenzene (*p*-NCB) (Wu et al., 2014), and orange IV azo dye ($C_{18}H_{14}N_3NaO_3S$) (Wang et al., 2015). The degradation rate of TCE increased by 17% when using the aged NZVI system coupled with *Schewanella* compared to that observed when using aged NZVI alone (Yang et al., 2017). While the removal rate of U(VI) under the NZVI-CNT condition was only 47% in a 120-min period, it improved to 97% when *Schewanella*-NZVI-CNT was employed (Xiang et al., 2020). These observations indicate that *Schewanella* could be an ideal activator to recover and improve the performance of aged NZVI/GO for the rapid removal of TBBPA.

To investigate TBBPA removal efficiency and clarify the underlying reactivation mechanism, and subsequently improve the performance of the aged NZVI/GO with the introduction of *Schewanella*, a reaction system where *Shewanella putrefaciens* CN32 coupled with aged NZVI/GO (aged NZVI/GO+

CN32) was developed. The TBBPA removal performance of this system was then investigated in several stages: (1) to evaluate the TBBPA removal efficiency of aged NZVI/GO + CN32, the kinetics of the TBBPA reaction were analyzed; (2) Tounderstand the mechanism of interaction of TBBPA in aged NZVI/GO + CN32, the elemental distribution, surface morphology, and secondary mineral species formed at the interface of aged NZVI/GO+CN32 after the reaction were analyzed by transmission electron microscopy (TEM) mapping, scanning electron microscopy (SEM), and X-ray diffraction (XRD), respectively; (3) To understand the related electron transfer mechanism and metabolic pathway of TBBPA in the aged NZVI/GO+CN32 system, the free radicals and intermediates were analyzed by using electron paramagnetic resonance spectrometry (EPR) and liquid chromatography-mass spectrometry (LC-MS).

5.3.2 Materials and methods

(1) Experimental reagents and microbial cultivation

TBBPA ($C_{15}H_{12}Br_4O_2$) was obtained from Shanghai Damas-beta Reagent Co., Ltd., China, while all other chemicals (analytical grade) were obtained from Shanghai Sinopharm Chemical Reagent Co. Ltd., China. *Shewanella putrefaciens* CN32 (CN32) was obtained from the China Marine Culture Collection.

The mixtures of 5 mL CN32 stock solution and 50 mL sterilised nutrient solution were prepared in 100-mL sterilised shake flasks for CN32 activation and enrichment. The solutions were then cultured in a 30 ℃ constant-temperature shaker at 200 rpm. The formulas for the nutrient solution for CN32 cultivation and TBBPA-removal batch experiments are shown as follows.

The nutrient solution used for CN32 cultivation contained 10.0 g peptone, 10.0 g NaCl, and 5.0 g yeast extract dissolved in 1 L Milli-Q water, the pH was adjusted to 6.9 – 7.1 by the 1 M NaOH. The nutrient solution was sterilized at 121 ℃ for 20 minutes. The 1-L nutrient solution used for TBBPA-removal experiments contained 0.460 g NH_4Cl, 0.225 g K_2HPO_4, 0.225 g KH_2PO_4, 0.117 g $MgSO_4 \cdot 7H_2O$, 0.225 g $(NH_4)_2SO_4$, 1% yeast extract, and 50 mM HEPES. The nutrient solution was sterilized at 121 ℃ for 20 minutes.

(2) Evaluation of the impact of the addition of CN32 to the aged NZVI and NZVI/GO systems on the removal of TBBPA

a. To study the effect of CN32 on the restoration of the performance of aged NZVI and aged NZVI/GO in terms of removal of TBBPA, six control experiments were performed using inactivated CN32, CN32, aged NZVI, aged NZVI/GO, aged NZVI + CN32, and aged NZVI/GO + CN32. The preparation methods for aged NZVI and aged NZVI/GO as well as the XRD patterns for both systems are shown as follows in and Figure 5.13.

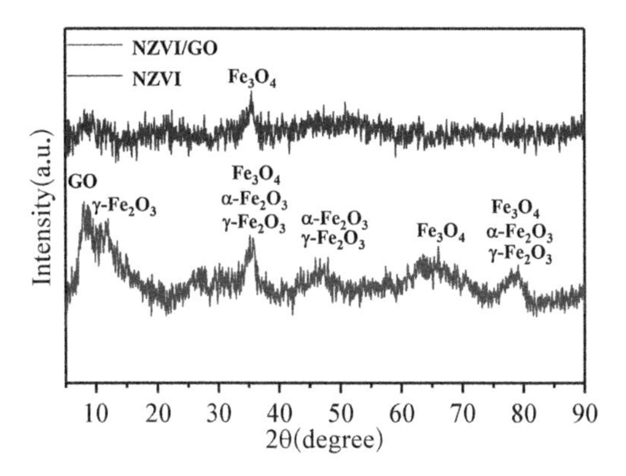

Figure 5.13 XRD patterns of the aged NZVI and NZVI/GO.

The preparation method of aged NZVI and aged NZVI/GO is as follows, 0.10 g fresh NZVI or 0.20 g NZVI/GO amendment was added in 100-mL shake flasks for treating 10.0 mg·L^{-1} TBBPA solution (100 mL) under pH = 7.0±0.5 condition. It was cultivated in a 30 ℃ constant-temperature shaker at 200 rpm. The mass ratio (r_m) of NZVI to GO in the NZVI/GO amendment was 1.0. Each batch experiment lasted 24 hours and a fresh TBBPA solution was used to replace the previous TBBPA solution for the next batch experiment. The batch experiments were performed at least 5 times to obtain the aged NZVI and aged NZVI/GO.

Following preparation, the aged NZVI prepared from 0.10 g fresh NZVI, 20.0 mM electron donor sodium lactate, and 100 mL of nutrient solution were added to 125-mL sealed and sterilised serum bottles. The initial concentrations of CN32 and TBBPA in the protocol for aged NZVI +CN32 were 1.25 g·L^{-1} and 10.0 mg·L^{-1}, respectively.

To maintain the iron content constant before aging, 0.20 g of fresh NZVI/GO was added during the preparation of aged NZVI/GO+CN32. The mass ratio (r_m) of NZVI to GO was 1.0. The same materials and methods used in the other protocols were also applied for NZVI + CN32. Following preparation, mixtures were cultivated in a 30 ℃ constant-temperature shaker at 200 rpm. The pH was 7.1±0.5, and three parallel experiments were performed. Water samples were collected at specific time intervals for the UV-Vis spectra of total Fe(II) as well as sorbed Fe(II) and the high-performance liquid chromatography (HPLC) analysis of TBBPA and related by-products.

b. To investigate the effect of the aged NZVI/GO dosage on the TBBPA removal kinetics in aged NZVI/GO + CN32 reaction system, four different concentrations of NZVI/GO (0.1 $g \cdot L^{-1}$, 0.5 $g \cdot L^{-1}$, 1.0 $g \cdot L^{-1}$, and 2.0 $g \cdot L^{-1}$) were prepared, along with sodium lactate (20.0 mM), CN32 solution (1.25 $g \cdot L^{-1}$), and a TBBPA solution (10.0 $mg \cdot L^{-1}$). The initial mass ratio (r_m) of NZVI to GO before aging was 1.0.

c. To investigate the effect of sodium lactate dosage on TBBPA removal kinetics in the aged NZVI/GO + CN32 reaction system, four different concentrations of sodium lactate (1.0 mM, 5.0 mM, 10.0 mM, and 20.0 mM), 1.0 $g \cdot L^{-1}$ aged NZVI/GO, CN32 solution (1.25 $g \cdot L^{-1}$), and a TBBPA solution (10.0 $mg \cdot L^{-1}$) were prepared. The r_m of NZVI to GO was 1.0. The three-parameter single-index pseudo-first-order kinetic model (Eq. (1)) (Xiang et al., 2021a) and the trial and error method was adopted as the curve-fitting approach.

$$C_t = C_{ultimate} + (C_0 - C_{ultimate}) \exp(-k_{obs} \cdot t). \tag{5-3}$$

Other materials and methods in experiments (b) and (c) were the same as those used in experiment (a).

(3) The analysis of secondary minerals and TBBPA intermediates formed during the reaction

Following the completion of the experiment (a), the filtered water samples were analyzed for TBBPA intermediates by LC-MS. Additionally, solid amendment samples were collected for SEM, TEM mapping, and XRD analysis. Furthermore, metabolic pathways and reaction mechanisms will be proposed based on the results of identified TBBPA intermediates, surface morphology, and the secondary mineral species formed at the interface of the aged NZVI/GO+CN32 amendment. Additional analysis of the four dosages of the aged NZVI/GO amendments in the experiment (b) was performed by EPR to provide additional evidence to support the proposed reaction mechanism.

(4) Analytical methods

TBBPA was determined by HPLC (Waters 2695, USA) equipped with a diode array detector (DAD) with a wavelength of 210 nm and a Water SunFire-C18 column (4.6×150 mm, 5 μm). The mobile phase was a mixture of methanol: water solution (80 : 20, v/v), and the flow rate was 1 $mL \cdot min^{-1}$. The intermediates of TBBPA were analyzed by LC-MS (TSQ Quantum Ultra, Thermo Fisher Scientific, USA) equipped with an Accucore-C18 column (2.1 mm× 100 mm, 2.6 μm). The wavelength was 210 nm and the column temperature was 25 ℃. The mobile phase was a mixture of acetonitrile and water solution with a gradient elution program and the flow rate was 0.6 $mL \cdot min^{-1}$. The HCl-extractable Fe(II) (sorbed Fe(II)) and dissolved Fe(II) were analyzed at 510 nm using ultraviolet spectrophotometry (TU-1901, Puxi, China). The surface morphology was investigated using a TEM (JEM-2100, JEOL, Japan), biological SEM (G300, ZEISS, Germany), and field emission SEM (FESEM) (S4800, Hitachi, Japan). The secondary mineral species were analyzed by using XRD (Advance D8, Bruker, Germany) at 40 kV voltage and 40 mA current using Cu Ka radiation ($\lambda = 1.5406$ Å) in the range (10°−90°) at a scanning rate of 5° min^{-1}. The semiquinone radical on aged NZVI/GO was analyzed by EPR (A300, Bruker, Germany).

5.3.3 Results and discussion

(1) Effect of CN32 on the removal kinetics of TBBPA in different reaction systems

The aged NZVI/GO+CN32 system was shown to be the most effective in the removal of TBBPA with a removal rate of 92% (Figure 5.14 (a)). This was significantly higher than the TBBPA removal obtained by aged NZVI + CN32

(64%), aged NZVI/GO (27%), CN32 (15%), and aged NZVI (6%). Compared to the aged NZVI+CN32 system, CN32 showed a much better synergistic effect on the aged NZVI/GO+CN32 system (Figure 5.14 (a)). In addition, the removal of TBBPA conformed to the pseudo-first-order kinetic model (Eq. (1)). The k_{obs} and residual TBBPA concentration $C_{ultimate}$ in the aged NZVI/GO+CN32 system were 0.52 day^{-1} and 1.25 $mg \cdot L^{-1}$, much higher and lower, respectively, than those in the aged NZVI+CN32 at 0.47 day^{-1} and 3.65 $mg \cdot L^{-1}$, the aged NZVI/GO at 0.45 day^{-1} and 7.28 $mg \cdot L^{-1}$, the CN32 at 0.44 day^{-1} and 8.58 $mg \cdot L^{-1}$, and the aged NZVI at 0.43 day^{-1} and 9.46 $mg \cdot L^{-1}$ respectively (Table 5.7). The above results demonstrated that CN32 could effectively decompose the surface iron oxide/hydroxide passivation layer to improve the performance of the aged NZVI/GO, achieving the highest reaction rate and best adsorption performance of the systems.

Table 5.7 Reaction parameters for TBBPA removal in CN32, aged NZVI, aged NZVI/GO, aged NZVI + CN32, and aged NZVI/GO + CN32 systems; as well the reaction parameters of TBBPA in aged NZVI/GO+CN32 reaction system under different concentrations of sodium lactate and aged NZVI/GO conditions.

	Protocol	k_{obs} (day^{-1})	C_u ($mg \cdot L^{-1}$)	R^2
aged NZVI	$C_{aged\ NZVI}$ = 1.0 $g \cdot L^{-1}$, 20.0 mM sodium lactate	0.43	9.46	0.99
CN32	C_{CN32} = 1.25 $g \cdot L^{-1}$, 20.0 mM sodium lactate	0.44	8.58	0.99
aged NZVI/GO	$C_{aged\ NZVI/GO}$ = 2.0 $g \cdot L^{-1}$, 20.0 mM sodium lactate	0.45	7.28	0.97
aged NZVI +CN32	$C_{aged\ NZVI}$ = 1.0 $g \cdot L^{-1}$, C_{CN32} = 1.25 $g \cdot L^{-1}$, 20.0 mM sodium lactate	0.47	3.65	0.96
aged NZVI/GO+ CN32	$C_{aged\ NZVI/GO}$ = 2.0 $g \cdot L^{-1}$, C_{CN32} = 1.25 $g \cdot L^{-1}$, 20.0 mM sodium lactate	0.52	1.25	0.97
aged NZVI/GO+ CN32	1.0 mM sodium lactate, $C_{aged\ NZVI/GO}$ = 2.0 $g \cdot L^{-1}$, $C_{CN\ 32}$ = 1.25 $g \cdot L^{-1}$	0.19	4.86	0.96
	5.0 mM sodium lactate, $C_{aged\ NZVI/GO}$ = 2.0 $g \cdot L^{-1}$, C_{CN32} = 1.25 $g \cdot L^{-1}$	0.47	4.78	0.97

(continued)

	Protocol	k_{obs} (day^{-1})	C_u ($mg \cdot L^{-1}$)	R^2
aged NZVI/GO+CN32	10.0 mM sodium lactate, $C_{aged\ NZVI/GO}$ = 2.0 $g \cdot L^{-1}$, C_{CN32} = 1.25 $g \cdot L^{-1}$	0.48	3.65	0.97
	20.0 mM sodium lactate, $C_{aged\ NZVI/GO}$ = 2.0 $g \cdot L^{-1}$, C_{CN32} = 1.25 $g \cdot L^{-1}$	0.52	1.25	0.99
aged NZVI/GO+CN32	$C_{aged\ NZVI/GO}$ = 0.1 $g \cdot L^{-1}$, C_{CN32} = 1.25 $g \cdot L^{-1}$, 20.0 mM sodium lactate	0.19	6.06	0.96
	$C_{aged\ NZVI/GO}$ = 0.5 $g \cdot L^{-1}$, C_{CN32} = 1.25 $g \cdot L^{-1}$, 20.0 mM sodium lactate	0.22	2.95	0.98
	$C_{aged\ NZVI/GO}$ = 1.0 $g \cdot L^{-1}$, C_{CN32} = 1.25 $g \cdot L^{-1}$, 20.0 mM sodium lactate	0.35	2.55	0.99
	$C_{aged\ NZVI/GO}$ = 2.0 $g \cdot L^{-1}$, C_{CN32} = 1.25 $g \cdot L^{-1}$, 20.0 mM sodium lactate	0.52	1.25	0.99

[a] The mass ratio of NZVI-to-GO was 1.0 and the initial concentration of TBBPA was 10.0 $mg \cdot L^{-1}$.

Furthermore, the total Fe(II) (0.44 mM) and sorbed Fe(II) (0.32 mM) in the aged NZVI/GO + CN32 system were both higher than the corresponding concentrations (0.21 mM, 0.14 mM) in the aged NZVI+CN32 system after the reaction (Figure 5.14 (b) and (c)). This was also true for the concentration of free Fe(II) ion (0.12 mM) in aged NZVI/GO+CN32, which was again higher than the observed concentration (0.07 mM) in aged NZVI + CN32. The percentage of sorbed Fe(II) on the interface of the aged NZVI/GO+CN32 amendment and the aged NZVI + CN32 amendment were 73% and 67%, respectively. This demonstrated that the sorbed Fe(II) was significantly greater than free Fe(II) ions. The results further demonstrated that *Shewanella* CN32 had a much better efficiency in decomposing the iron oxide passivation layer of aged NZVI/GO, while also releasing more Fe^{2+} ions (Figure 5.14 (b) and (c)). This was due to the metabolization of sodium lactate by She*wanella* CN32, which also produced electrons, resulting in a micro-electrolysis cycle for corrosion of the iron oxide passivation layer. The results revealed that the produced electrons can transfer much faster to the surface layer of the aged NZVI/GO + CN32 than the aged NZVI+CN32.

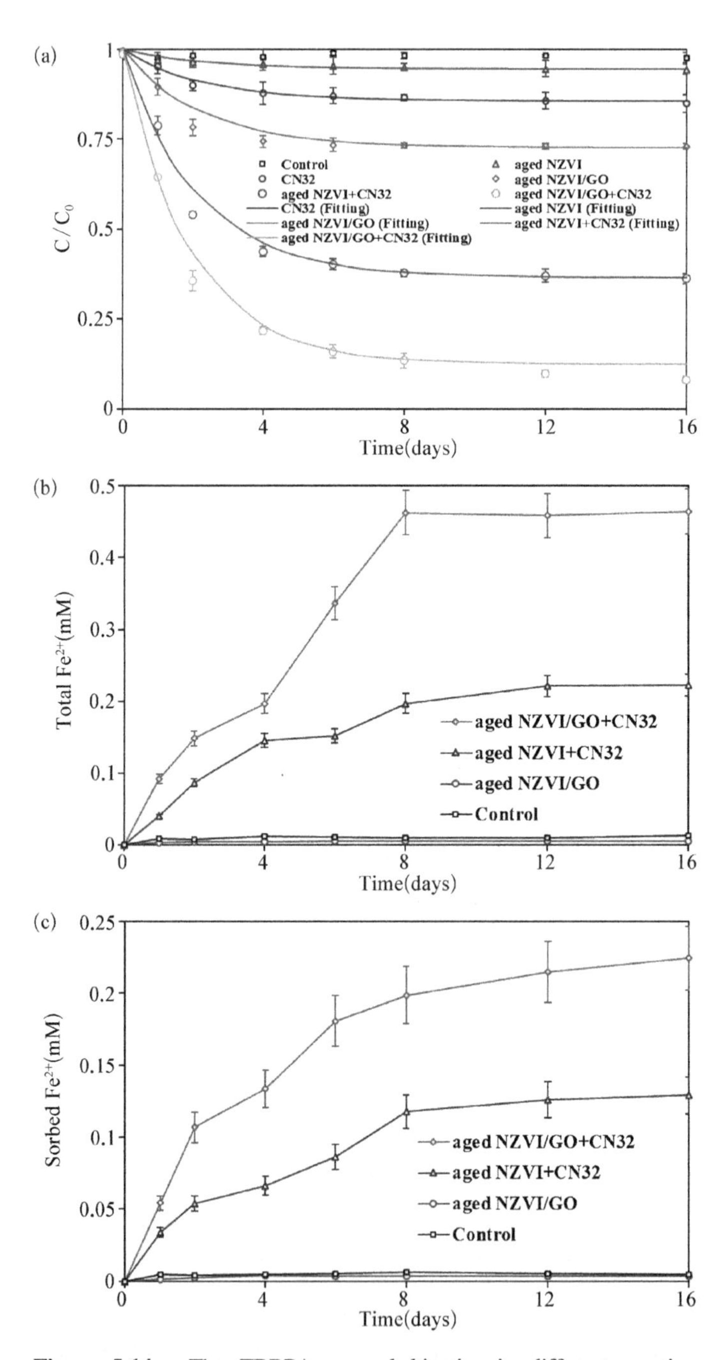

Figure 5.14 The TBBPA-removal kinetics in different reaction systems (a); the concentration of total Fe(Ⅱ) and sorbed Fe(Ⅱ) in different reaction systems (b)-(c); CN32 represents the *Shewanella putrefaciens* CN32; Control was the inactivated CN32.

(2) Effect of NZVI/GO amendment and electron donor concentrations on the removal kinetics of TBBPA

The results demonstrated that the effect of the amendment and electron donor concentrations on TBBPA removal was significant and effective; 89% and 88% of TBBPA removal was achieved in the aged NZVI/GO+CN32 system when the concentration of NZVI/GO amendment increased from 0.1 $g \cdot L^{-1}$ to 2.0 $g \cdot L^{-1}$ and the dosage of electron donor sodium lactate increased from 1.0 mM to 20 mM, respectively (Figure 5.15). The k_{obs} and $C_{ultimate}$ in the aged NZVI/GO+CN32

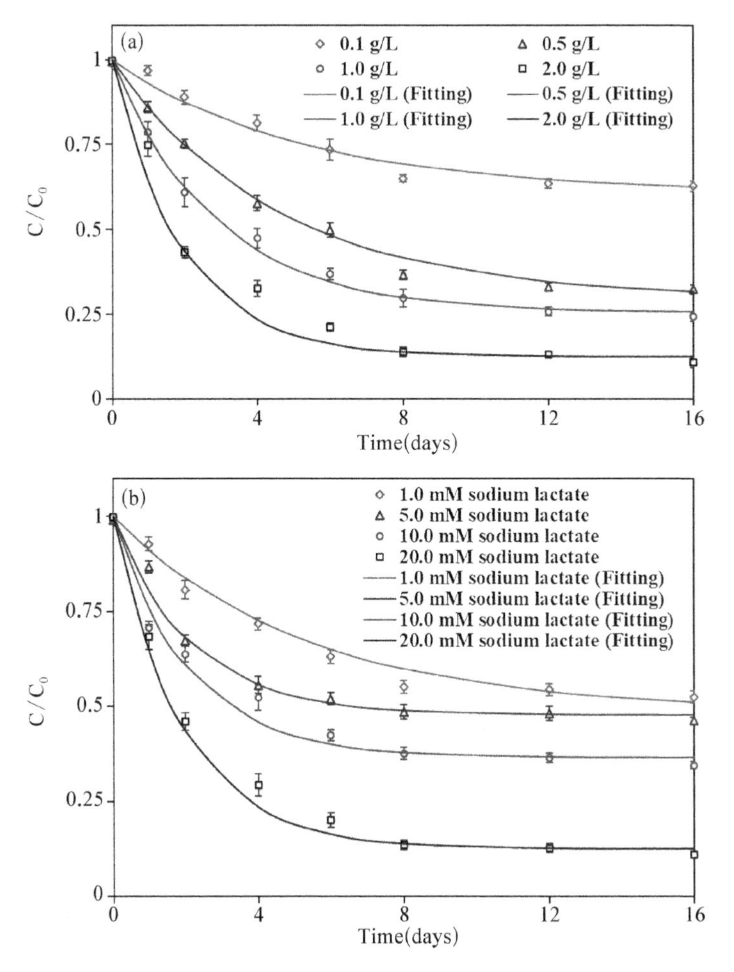

Figure 5.15 The effect of initial concentrations of aged NZVI/GO (a) and sodium lactate (b) on the TBBPA-removal kinetics in the aged NZVI/GO+CN32 system.

system were 0.19 day^{-1} and 4.86 $mg \cdot L^{-1}$ under 1.0 mM sodium lactate condition (Table 5.7). When the sodium lactate concentration increased to 20.0 mM, the k_{obs} increased to 0.52 day^{-1} and the $C_{ultimate}$ decreased to 1.25 $mg \cdot L^{-1}$ (Table 5.7). The resulting reaction and adsorption performance increased by 174% and 289%, respectively.

For the aged NZVI/GO+CN32 system under the amendment condition of 0.1 $g \cdot L^{-1}$, the k_{obs} and $C_{ultimate}$ were 0.19 day^{-1} and 6.06 $mg \cdot L^{-1}$. Increasing the amendment to 2.0 $g \cdot L^{-1}$, again resulted in an increment of the k_{obs} to 0.52 day^{-1} and a decrease in the $C_{ultimate}$ to 1.25 $mg \cdot L^{-1}$ (Table 5.7). The reaction performance and the adsorption performance increased by 174% and 385%, respectively. The results demonstrated that increasing the dosage of the amendment and increasing the dosage of the electron donor sodium lactate both improved the TBBPA reaction rate (Table 5.7). The relevant reaction mechanism is discussed in following Section (4) in this study. Furthermore, the comparison of the TBBPA removal rate in the current study with the previously reported work was provided in Table 5.8. The results demonstrated that the removal rate of TBBPA in our study was significantly greater than that of the previously reported work.

Table 5.8 Comparison of TBBPA removal rate in current study with previously reported work.

Reaction system	k_{obs} (day^{-1})	TBBPA initial concentration	Reference
aged NZVI/GO+CN32	0.52	10 $mg \cdot L^{-1}$	This study
Pb/Fe-microorganism	0.46	15 μM	(Lin et al., 2020)
Pycnoporus sanguineus	0.33	5 $mg \cdot L^{-1}$	(Feng et al., 2019)
Ochrobactrum sp.	0.25	10 $mg \cdot L^{-1}$	(An et al., 2011)
Pseudoalteromonas sp.	0.15	50 $mg \cdot L^{-1}$	(Gu et al., 2018)
nitrifying activated sludge	0.13	1 $mg \cdot L^{-1}$	(Li et al., 2015)
JXS-2-02	0.08	10 $mg \cdot L^{-1}$	(Peng et al., 2013)
Sludge & SMC	0.07	100 $mg \cdot kg^{-1}$	(Yang et al., 2017)
Fe/Ni-microorganism	0.06	2 $mg \cdot L^{-1}$	(Peng et al., 2017)
CNT/α-FeOOH-CN32	0.03	20 $mg \cdot L^{-1}$	(Li et al., 2021)

(3) Secondary minerals formed at the interface of aged NZVI/GO+CN32

The XRD results of aged NZVI and aged NZVI/GO are provided in Figure 5.13, where Fe_3O_4 was the dominant mineral form of iron oxide found in aged NZVI, while Fe_3O_4, α-Fe_2O_3, and γ-Fe_2O_3 were the main iron oxides minerals observed in aged NZVI/GO. The SEM results showed that secondary minerals (Figure 5.16 b) and CN32 (Figure 5.16 (c) and (e)) covered the surface of aged NZVI/GO+CN32 after the reaction. In addition, the SEM results showed that the typical morphologies of secondary minerals appeared in the shape of lumps and strips (Figure 5.16 (d) and (f)).

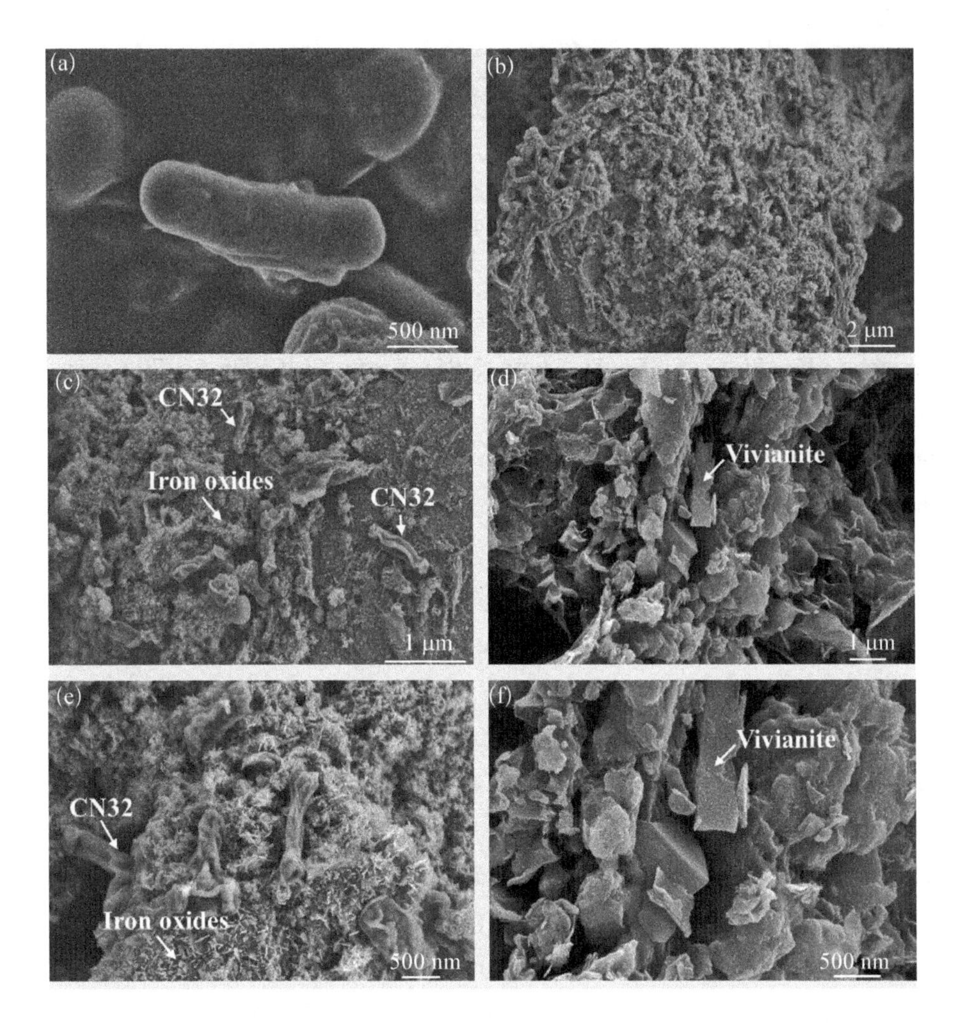

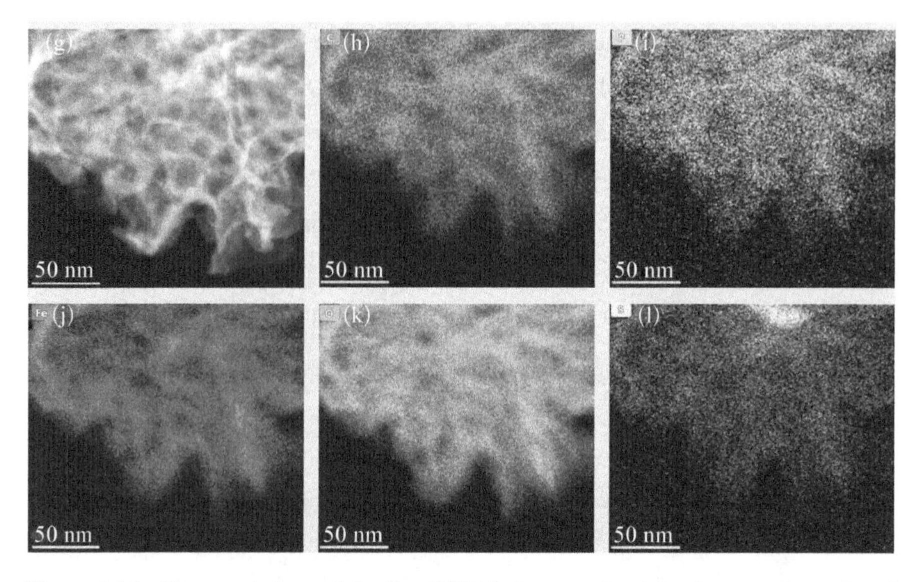

Figure 5.16 The morphology of the free CN32 before reaction (a); the secondary mineral and CN32 at the interface of aged NZVI/GO + CN32 after the reaction (b)-(f); TEM element mapping (C, P, Fe, O, and S) of aged NZVI/GO (g)-(l); CN32 represents the *Shewanella putrefaciens* CN32.

The morphological features indicated the formation of structural Fe(II) vivianite ($Fe_3(PO_4)_2 \cdot 8H_2O$), with XRD spectra (JCPDS # 3-70) further confirming this (Figure 5.17 (a)). The formation of vivianite was due to the presence of PO_4^{3-} in the nutrient solution that reacted with the Fe^{2+} released from the aged NZVI/GO. In addition, based on XRD results, SO_4^{2-} type green rust (JCPDS#13-90) ($Fe_{3.6}Fe_{0.9}(O, OH, SO_4)_9$), Fe_3O_4 (JCPDS #1-111), α-Fe_2O_3 (JCPDS#5-637) and γ-Fe_2O_3 (JCPDS #4-755) were also produced in the aged NZVI/GO+CN32 system after the reaction (Figure 5.17 (a)).

In contrast, no vivianite ($Fe_3(PO_4)_2 \cdot 8H_2O$) was formed in the aged NZVI+CN32 system. Only the SO_4^{2-} type green rust (JCPDS#13-90), the CO_3^{2-} type green rust ($Fe_6(OH)_{12}CO_3$) (JCPDS#46-98), and α-Fe_2O_3 (JCPDS# 5-637) were produced (Figure 5.17 (a)). These results demonstrated that GO accelerated the reduction of dissimilatory iron and the formation of Fe(II) vivianite crystals (Figures 5.16 and 5.17 (a)). This was consistent with previously reported findings of graphite accelerated reduction of dissimilatory iron (α-Fe_2O_3) and vivianite crystal formation (Wu et al., 2021). The observed levels of P

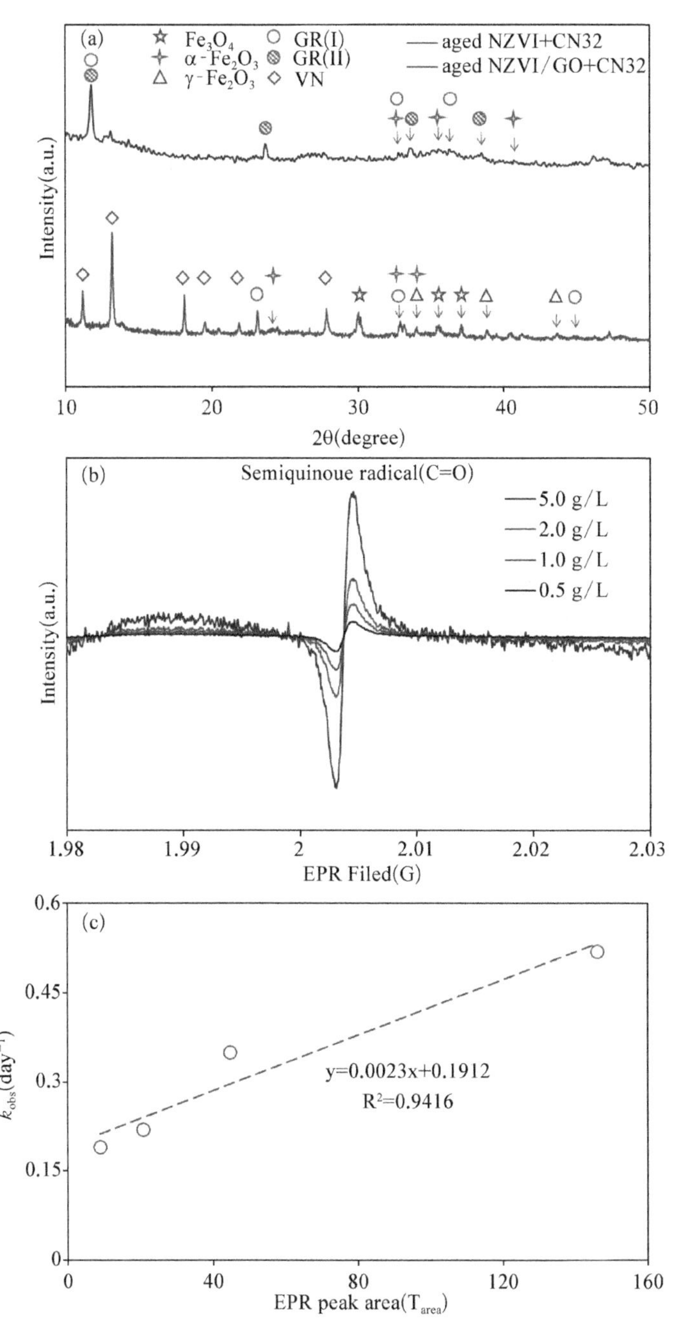

Figure 5.17 The secondary mineral species in the interface of amendment after the reaction (a); VN represents the vivianite ($Fe_3(PO_4)_2 \cdot 8H_2O$) (JCPDS#3-70); GR(I) and GR(II) represents the SO_4^{2-} type green rust (JCPDS#13-90) ($Fe_{3.6}Fe_{0.9}(O,OH,SO_4)_9$) and CO_3^{2-} type green rust ($Fe_6(OH)_{12}CO_3$) (JCPDS#46-98), respectively; the intensity of the semiquinone radical (C = O) for different dosages of aged NZVI/GO under the electron paramagnetic resonance (EPR) filed (b); the correlation coefficient between the k_{obs} and the intensity of the C=O (c).

(Figure 5.16 (i)) were higher than those of elemental S (Figure 5.16 (l)), demonstrating that the formation of vivianite was greater than that of the SO_4^{2-} type green rust. Furthermore, TEM element mapping (50 nm scale) showed that the P, Fe, and S content in the peak tops and slopes of the GO interface were greater than those in the peak valleys of the GO interface. The elemental distribution in the peak tops and slopes of the GO interface appeared to be more uniform (Figures 5.16 (g)-(l)).

(4) The reaction mechanism and metabolic pathway in the aged NZVI/GO+CN32 system

Figure 5.17 (b) showed an electron paramagnetic resonance (EPR) G value between 1.996 and 2.004, corresponding to the typical range of G-values observed for semiquinone radicals (C = O) (Li et al., 2021), which aligns with previous research demonstrating that C = O could accelerate electron transfer (Li et al., 2021). In this study, there is an obvious linear relationship ($R^2 = 0.94$) between the k_{obs} of TBBPA (Figure 5.15 (a)) and the EPR peak area (T_{area}) of C = O (Figure 5.17 (b)) under four amendment concentration conditions (Figure 5.17 (c)). Therefore, the improvement of k_{obs}(from 0.19 to 0.52 day^{-1}) (Figure 5.15 (a)) was attributed to the fact that increasing the amendment concentrations (from 0.1 $g \cdot L^{-1}$ to 2.0 $g \cdot L^{-1}$) provided higher levels of Fe^0, iron oxides, and C = O (Figure 5.17 (b)). This reduction of Fe^0 combined with the decomposition of iron oxides by CN32 ensured an increased yield of Fe^{2+} ions and electrons (Figure 5.14 (b)); after which, the presence of C = O radicals further accelerated TBBPA removal through faster electron transfer (Li et al., 2021). By increasing the sodium lactate dose (from 1.0 mM to 20.0 mM), higher levels of carbon resource were provided to the system, leading to the improved growth of CN32, which in turn further accelerated the release of Fe^{2+} and electrons from the Fe^0 and iron oxides, thus increasing the k_{obs} of TBBPA (Figure 5.15 (b)). The secondary mineral vivianite (Figures 5.16 and 5.17 (a)) produced by the biological reduction of iron oxide, also acts as a strong reducing agent (Bae et al., 2017; Bae and Lee, 2012), further accelerating the degradation of TBBPA.

Based on the results of LC-MS and previous reports (Ding et al., 2013; Fan et al., 2017; Kang et al., 2018; Macêdo et al., 2021; Wang et al., 2021), the

TBBPA degradation pathway in the aged NZVI/GO+CN32 system is shown in Figure 5.18.

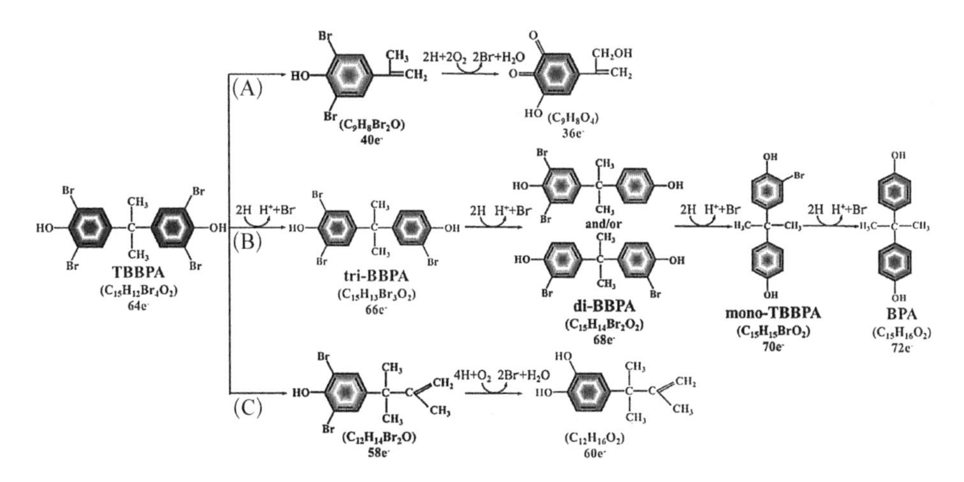

Figure 5.18 The debromination path of TBBPA in aged NZVI/GO+CN32 system based on the previous reports (Ding et al., 2013; Fan et al., 2017; Kang et al., 2018; Macêdo et al., 2021; Wang et al., 2021) and our study. The red marks represent the intermediates that were detected by LC-MS in our study. Electron equivalents: $C_{15}H_{12}Br_4O_2 + 28H_2O \rightarrow 15CO_2 + 64H + 4HBr$, where 64H represents an electron carrier with 64 electron equivalents.

In pathway (A), the central carbon bond of TBBPA is broken to obtain the dibrominated product $C_9H_8Br_2O$, at which point a process of debromination and hydroxylation replaces the bromine atom to form the intermediate $C_9H_8O_4$ (Ding et al., 2013). Pathway (B) is a stepwise debromination process. NZVI easily adsorbed TBBPA, where each C-Br bond is gradually broken, and Br atoms are replaced by H atoms and generate the final product bisphenol A ($C_{15}H_{16}O_2$) (Kang et al., 2018; Li et al., 2016; Luo et al., 2010). In this process, the debromination pathway was TBBPA → tribromobisphenol A (tri-BBPA) → dibromobisphenol A (di-BBPA) → monobromobisphenol A (mono-BBPA) → bisphenol A (BPA) (Kang et al., 2018; Wang et al., 2021). Pathway (C) is the biodegradation process of TBBPA, in which the intermediate $C_{12}H_{14}Br_2O$ is generated through two pathways: (Ⅰ) beta-scission occurs in TBBPA and (Ⅱ) debromination, decomposition, and dehydrogenation reactions (Fan et al., 2017; Macêdo et al., 2021). Finally, the final product, $C_{12}H_{16}O_2$, forms through the debromination and hydroxylation of $C_{12}H_{14}Br_2O$.

Based on the results shown in Figures 5.15 – 5.18, a reaction mechanism for the removal of TBBPA in the aged NZVI/GO + CN32 system is proposed and shown in Figure 5.19. As an electron shuttle, GO adsorbed CN32 and TBBPA on the surface. The electron donor sodium lactate was oxidised by CN32 through its characteristic enzymatic reaction, and electrons were generated during the oxidation process. Electrons were further transferred to GO through the reaction chain and then to Fe(III) iron oxides (Fe_3O_4, α-Fe_2O_3, and γ-Fe_2O_3) on the surface of aged NZVI/GO (Figure 5.13), which was reduced to the adsorbed structural Fe(II) vivianite crystal and the SO_4^{2-} type green rust.

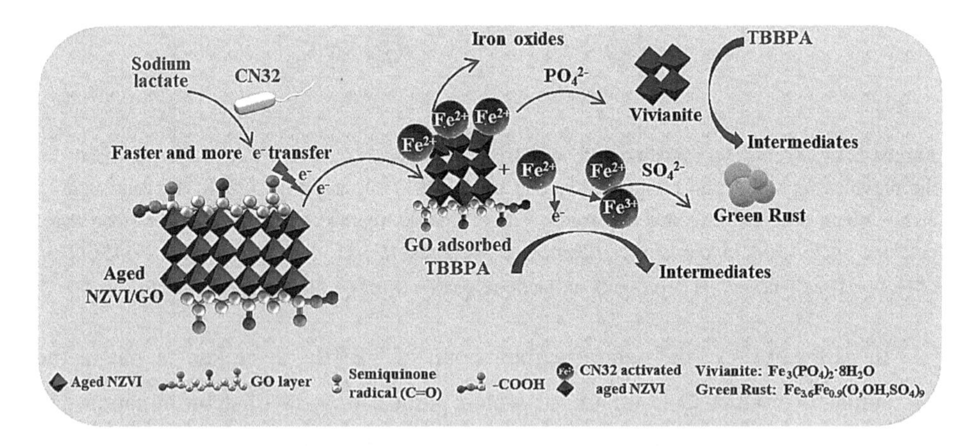

Figure 5.19 Proposed reaction mechanism of TBBPA in the aged NZVI/GO+CN32 system; CN32 represents the *Shewanella putrefaciens* CN32.

5.3.4 Conclusion of this case study

The present study demonstrated that CN32 significantly improved the performance of aged NZVI/GO. The percentage (92%) of TBBPA removal in the aged NZVI/GO+CN32 system increased by 240% compared to that observed with the aged NZVI/GO system (27%). Most (75%) of the Fe (II) present at the interface of aged NZVI/GO+CN32 was found to be the sorbed Fe(II). Compared to the aged NZVI, CN32 showed much better activation performance in destroying the iron oxide passivation layer of aged NZVI/GO and releasing more Fe^{2+} ions. The results of LC-MS further demonstrated that the reaction path of TBBPA in the aged NZVI/GO + CN32 system included two debromination and hydroxylation processes, and one step-by-step debromination process. Vivianite and SO_4^{2-} type green rust secondary minerals were formed during the reaction process. GO accelerated dissimilatory iron reduction and structural Fe (II) vivianite crystal formation, and vivianite further improved the debromination of TBBPA. The semiquinone radical (C=O) existing at the interface of aged NZVI/GO also has a strong mediating ability, thus effectively mediating electron transfer and recovering the performance of aged NZVI/GO by CN32. In conclusion, the main reaction process and the TBBPA removal mechanisms in the aged NZVI/GO + CN32 reaction system were shown to be CN32-induced iron oxide reduction and NZVI exposure, C = O mediating electron transfer, and vivianite-induced TBBPA debromination. In future applications, the aged NZVI/GO should be more conveniently reactivated by CN32 and further efficiently used in the remediation of soil, groundwater, and lakes.

5.4 Case study for PFOA transport and reaction in microbe-Fe_xS_y interaction media

5.4.1 Research background

Per- and poly-fluoroalkyl substances (PFAS) find extensive applications in disposable food packaging (Schaider et al., 2017), non-stick cooking utensils (Sunderland et al., 2019), and aqueous film-forming foam (Trang et al., 2022), resulting in ubiquitous contamination of environmental media (Lim, 2019). Among these, perfluorooctanoic acid (PFOA) emerges as one of the most prevalent PFAS, characterized by its capacity for long-range transport, high persistence, toxicity, and bioaccumulation potential (Teng et al., 2024). Due to the great bond energy of C-F bonds, PFOA exhibits excellent high surface activity, hydrophobicity, and robust thermal and chemical stability (Liu et al., 2017). This compound has been consistently detected in various matrices, including biota (Zhang et al., 2019), atmosphere (Cousins et al., 2022), soil (Galloway et al., 2020), and groundwater (Xiao et al., 2017). The PFOA poses significant risks to human health, disrupting the reproductive, nervous, and immune systems, and even elevating cancer risks (Wang et al., 2016). Acknowledging its hazards, the US Environmental Protection Agency has stipulated a novel drinking water concentration limit for PFOA at levels below 4.0 $ng \cdot L^{-1}$ (USEPA, 2023). At present, PFOA has been included in the list of persistent organic pollutants (Falk et al., 2019; Lyu et al., 2020a). Therefore, comprehending the transport behavior and its underlying mechanisms of PFOA in soil and groundwater has garnered widespread attention, as it informs the regulations for contamination control and prevention.

Currently, the transport behavior of PFOA in quartz sand porous media has been the subject of several researches. These investigations have revealed that various chemical conditions of the solution, such as ionic strength and the type of cations present (Li et al., 2021; Lv et al., 2018), water saturation (S_w) (Brusseau et al., 2019; Lyu et al., 2018), and the chemical and physical characteristics of the media (organic matter and Al/Fe oxides) (Li et al., 2018; Lyu et al., 2019) collectively exerted a considerable effect on the transport and fate behavior of PFOA. Particularly noteworthy was the role played by mineral types (e.g. SiO_2, Al_2O_3, $Fe(OH)_3/Al(OH)_3$, and iron-sulfur minerals (Fe_xS_y)) and their content, which substantively impacted the transport and fate of PFOA. Hellsing et al. found that compared to SiO_2(with negative charges), Al_2O_3(with positive charges) had better retardation effects on PFOA transport, suggesting that electrostatic interaction affected PFOA transport behavior (Hellsing et al., 2016). Besides, Lyu et al. explored the effect of $Fe(OH)_3/Al(OH)_3$ minerals on the transport behavior of PFOA in saturated porous media. Their investigation unveiled a substantial retardation effect ($R = 1.28-5.58$) imposed by $Fe(OH)_3/Al(OH)_3$ on PFOA transport (Lyu et al., 2020b).

Moreover, studies have substantiated the capability of microorganisms to degrade PFOA through defluorination. Concurrently, it has been observed that PFOA can also impede the expression of specific genes within microorganisms. PFOA has the potential to hinder gene expression associated with various processes, including amino acid metabolism, energy production, and transformation within sludge microorganisms (Huang et al., 2022). Whereas, Fe_xS_y (e.g. pyrite

(FeS_2) and pyrrhotite ($Fe_{1-n}S$)) emerges with wide distribution in the Earth's crust, exhibiting susceptibility to weathering processes (Manceau et al., 2018; Saladino et al., 2008). In the realistic environment, the coexistence of Fe_xS_y and microorganisms is a common phenomenon, often leading to intricate and multifaceted interactions. Notably, Fe_xS_y functions as a "nutrient", offering trace elements that sustain microorganisms. In a symbiotic feedback loop, microorganisms, in turn, contribute to the transformation and regeneration of minerals (liquefied minerals) (Dong et al., 2022). Currently, a review of the existing literature above showed that studies about the transport and fate behavior of PFOA in the media accounting for microbe-mineral (e.g. Fe_xS_y) interaction are not enough.

Furthermore, research findings have demonstrated that the release of SO_4^{2-} (Yang et al., 2015), NO_3^-(Amano et al., 2016), and HCO_3^-(Qiu et al., 2022) ions were a common occurrence in areas associated with chemical plants, agricultural activities, and mining operations. The coastal regions adjoining seawater environments typically featured the presence of Cl^--Na^+ ions (Miao et al., 2021). Given these circumstances, PFOA, in conjunction with various ions, is prone to infiltrating both soil and groundwater, thereby engendering diverse PFOA-ions occurrence environments. However, the attenuation mechanism governing PFOA within the context of various PFOA-ions occurrence environments, especially in conjunction with microbe-mineral interactions, remains unclear.

Therefore, it is particularly necessary to explore the difference in effects between alone microbe or Fe_xS_y media and microbe-Fe_xS_y interaction media on PFOA and the specific difference in effects of various PFOA-ions occurrence environments on PFOA considering the microbe-Fe_xS_y interaction. The purposes of this study are: (i) to preliminarily explore the effects of microbe-Fe_xS_y interaction on the transport and fate behavior of PFOA in porous media; (ii) to assess the effect difference of different PFOA-ions occurrence environments on PFOA transport and fate under microbe-Fe_xS_y interaction condition; (iii) to employ mathematical models for simulation and validation of experimental data, thereby contributing to an enhanced understanding of changes in the transport and fate behavior of PFOA.

5.4.2 Materials and methods

(1) Experiment reagents

PFOA ($C_8HO_2F_{15}$, 98%) was obtained from the Bidepharmatech. The elemental sulfur (S), $Na_2S_2O_3 \cdot 5H_2O$, $FeSO_4 \cdot 7H_2O$, and quartz sand (20 mesh – 40 mesh, 1.65 $g \cdot cm^{-3}$) were obtained from Shanghai Titan Scientific Co., Ltd. Other chemicals (analytical grade) were all obtained from Shanghai Sinopharm Chemical Reagent Co. Ltd., China.

(2) Construction of microbe-Fe_xS_y interaction system

The interaction system setup consisted of microorganisms, Fe_xS_y-coated quartz sand, and a one-dimensional reaction column, as illustrated in Figure 5.20. The column was initially filled with Fe_xS_y-coated quartz sand and then infused with a mixed microorganism solution (2.16 $g \cdot L^{-1}$) at a flow rate of 0.5 $mL \cdot min^{-1}$ for 2 hours. Subsequently, the column was sealed at both ends and placed within an incubator set at 25℃ for 24 hours.

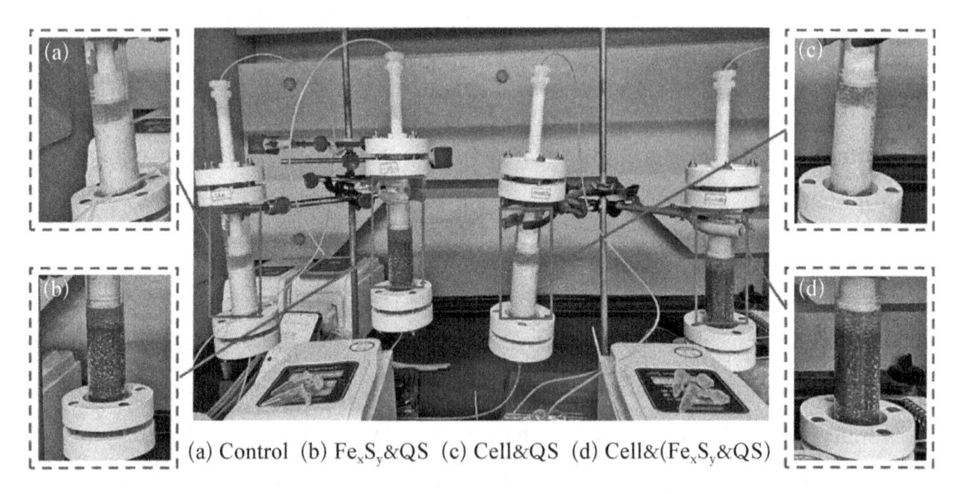

Figure 5.20 The application test device of one-dimensional packing column; type-a to type-d represent the Control, Fe_xS_y&QS, Cell&QS and Cell&(Fe_xS_y&QS) column, respectively. It includes the acrylic columns (Φ 2.5 cm, length 20.0 cm), porous medium, glass beads (Φ 5.0 cm, 11.0 g) and 100 mesh nylon filter cloth. Glass beads (Φ 5.0 mm, 11.0 g) were installed in the top and bottom of the column; porous medium (Φ 0.43 mm – 0.85 mm, 45.0 g) were installed in the middle of the column. The media in Control was replaced by clean quartz sand.

Detailed information about the preparation of Fe_xS_y(X-ray diffraction (XRD) and X-ray photoelectron spectroscopy (XPS) confirmed that the main composition was pyrite (FeS_2) and pyrrhotite ($Fe_{1-n}S$))-coated quartz sand was provided as follows. To obtain iron sulfur mineral Fe_xS_y-coated quartz sand, 0.04 mol of $FeSO_4 \cdot 7H_2O$, 0.04 mol of elemental S, and 0.04 mol of $Na_2S_2O_3 \cdot 5H_2O$ were added to the inner tank of the reactor containing 40.0 g of clean quartz sand (20 – 40 mesh). 120 mL of ultrapure water was added and stirred at room temperature for 30 min, then placed in a 200 ℃ oven for 24 hours, and cooled to room temperature. Then, clean it twice with ultrapure water and dry it in a vacuum drying oven at 60 ℃ for 5 hours (Yu. et al., 2020).

After the preparation of Fe_xS_y(pyrite (FeS_2) and pyrrhotite ($Fe_{1-n}S$))-coated quartz sand was completed, we store them in a nitrogen atmosphere. The microorganism employed in this study was adopted from our previous research (Wang et al., 2020) and underwent a domestication process utilizing PFOA. The target concentration of PFOA during domestication was 5 $mg \cdot L^{-1}$, which was continuously domesticated for two months, and then stored in a 25 ℃ incubator. The specific domestication steps are shown as follows.

Microorganisms, capable of biodegrading 2,4-DCP was used in previous work (Wang et al., 2020), it was sampled (500 mL) and inoculated in 2 200-mL filtering flasks with 200-mL headspace under not strictly anaerobic conditions. The flasks were sealed with Teflon-lined rubber septa with a pore (Φ 1.0 mm) and stored in an incubator at a constant temperature of 25 ℃. Every week, 1 000 mL of the culture solution was removed from the flasks, and the same volume of fresh nutrient solution (Ma and Wu, 2007) was added to each flask, and 20.0 mL PFOA solution (500 $mg \cdot L^{-1}$) was injected into each flask. Two months later, the domestication was finished and the flasks were kept in the dark at 25 ℃. Subsequently, in order to keep the activity of microorganism, once a month, 1 000 mL liquid from each culture flask was replaced with a fresh nutrient solution of the same volume to replenish the substrate nutrition, and 20.0 mL PFOA solution was injected. Before use, domesticate with nutrient solution and measure microbial concentration. Take microbial supernatant for the experiment.

The formula of the nutrient solution is as follows. The 1-L nutrient solution used to support microorganism cultivation contained 0.32 g NH_4Cl, 0.5 g KH_2PO_4, 0.1 g $CaCl_2$, 0.4 g $MgSO_4 \cdot 7H_2O$, 1.2 g $NaHCO_3$, 1.0 mL 10% yeast extract, 0.72 mL 60% sodium lactate, and 6.7 mL trace-element solution. The trace-element solution is 0.4 g $FeSO_4 \cdot 7H_2O$, 0.32 g $CoCl_2 \cdot 6H_2O$, 0.01 g H_3BO_3, 0.2 g $ZnSO_4 \cdot 7H_2O$, 0.04 g $CuSO_4 \cdot 5H_2O$, 0.2 g $NiSO_4 \cdot 6H_2O$, 0.1 g $MnCl_2 \cdot 4H_2O$, and 0.025 g $(NH_4)_6Mo_7O_{24} \cdot 4H_2O$ dissolved in 1 L ultrapure water (Ma and Wu, 2007).

(3) PFOA cross-media transport and fate behavior in the microbe-Fe_xS_y interaction media

a. To study the difference in effects between alone microbe/Fe_xS_y and microbe-Fe_xS_y interaction on the transport and fate behavior of PFOA, the Control (Quartz sand (QS) media), alone microbe (cell-coated quartz sand (Cell&QS) media), alone Fe_xS_y(Fe_xS_y-coated quartz sand (Fe_xS_y&QS) media) and microbe-Fe_xS_y interaction (microbe-Fe_xS_y coated quartz sand (Cell&(Fe_xS_y&QS)) media) systems were constructed as follows.

The constructed four-column systems include the acrylic columns (Φ 2.5 cm, length 20.0 cm), glass beads (Φ 5.0 mm), 100 mesh nylon filter cloth and Quartz sand (QS) media, cell-coated quartz sand (Cell&QS) media, Fe_xS_y-coated quartz sand (Fe_xS_y&QS) media and microbe-Fe_xS_y coated quartz sand (Cell&(Fe_xS_y& QS)) media. Fill the top and bottom of the column with glass beads (11.0 g), then place nylon filter cloth, and the clean quartz sand (45.0 g) was placed in the middle of the column for the Control system. Fe_xS_y-coated quartz sand (45.0 g) for the Fe_xS_y&QS system. For Cell&QS system, placed clean quartz sand (45.0 g) in the middle of the column, then pumped into microorganisms solution (2.16 $g \cdot L^{-1}$) at 0.5 $mL \cdot min^{-1}$ for 2 hours, then sealed both ends of the column, and placed it in a 25 ℃ incubator for the microorganisms loading and lasted 24 hours. In Cell& (Fe_xS_y&QS) system, Fe_xS_y-coated quartz sand (45.0 g) was placed in the middle of column, then the same steps were taken as above.

The concentration C_{PFOA} of pollution source was 1 $mg \cdot L^{-1}$. PFOA solution and nutrient solution were pumped into each column from the bottom for 72 hours (about 33 PVs − 45 PVs) at a flow rate of 0.16 $mL \cdot min^{-1}$. Ultrapure water was introduced into the column for 12 hours to elute PFOA, and the effluent was collected to obtain breakthrough curves and reveal the changes in PFOA transport and fate behavior. The column experiments were performed in duplicates.

b. To investigate the specific difference in effects of PFOA-ions occurrence on the transport and fate behavior of PFOA, PFOA-Cl^--Na^+, PFOA-HCO_3^-, PFOA-NO_3^-, and PFOA-SO_4^{2-} occurrence environments were simulated. The microbe-Fe_xS_y interaction system was employed. The solutions of 0.1 $mol \cdot L^{-1}$ NaCl, $NaHCO_3$, $NaNO_3$, and 0.05 $mol \cdot L^{-1}$ Na_2SO_4 were used to mix with 1.0 $mg \cdot L^{-1}$ of PFOA and nutrient solution for simulating different types of PFOA-ions occurrence environments. The effluent was collected to obtain breakthrough curves and reveal the changes in PFOA transport and fate behavior. The same other materials and methods were used in experiment (a). The column experiments were performed in duplicates.

c. To quantitatively analyze the difference in effects of the above interaction media types and PFOA-ions occurrence types on PFOA transport and fate behavior through changes in parameters, the multi-process microbe-Fe_xS_y interaction model was constructed for simulating the transport and fate behavior of PFOA. Furthermore, to obtain the PFOA spatial distribution and avoid error in direct testing, the accurate model by PFOA and tracer multivariate coupling of calibration for achieving synchronous inversion of PFOA transport behavior in water solution and distribution in solid spatial media. The accelerated column experiments for tracer in the Control, alone microbe, alone Fe_xS_y, and microbe-Fe_xS_y interaction media were performed. The flow rate was adjusted to 3.0 $mL \cdot min^{-1}$ to quickly obtain the dispersion coefficient without affecting the accuracy of the dispersion coefficient. Pumping 10.0 $g \cdot L^{-1}$ KCl solution into the column at 3.0 $mL \cdot min^{-1}$ for 50.0 minutes (5 PVs). Then, ultrapure water was pumped to wash off residual KCl for 50.0 minutes, and the effluent Cl^- was measured to obtain breakthrough curves. The other experimental steps were the same as (a).

(4) Analysis of iron-sulfur/oxide minerals transformation and PFOA reaction pathways

When experiment (b) finished, the solid samples were collected for analysis of mineral species and surface morphology by scanning electron microscope (SEM), XPS, and XRD. To obtain the PFOA intermediates in microbe-Fe_xS_y interaction systems under four types of PFOA-ions (PFOA-Cl^--Na^+, PFOA-HCO_3^-, PFOA-NO_3^-, and PFOA-SO_4^{2-}) occurrence conditions, batch experiments were performed and lasted 72 hours. In these experiments, the concentration C_{PFOA} of pollution source was set at 1.0 $mg \cdot L^{-1}$, while both microbe and Fe_xS_y were dosed at 2.16 $g \cdot L^{-1}$. Then the water samples were collected for intermediates analysis by liquid chromatograph mass spectrometer (LC-MS). The above analytical methods were provided as follows.

A liquid chromatograph mass spectrometer (LC-MS, Waters 1 Class, Sciex QTRAP 6500+) was used to detect the PFOA concentration quantitatively, and the Limits of Quantification (LOQ) and Limits of Detection (LOD) for PFOA were 5 $ng \cdot L^{-1}$ and 1 $ng \cdot L^{-1}$, respectively. The chromatographic column model used in LC-MS is Phenomenex 4 um Fusion-RP 80A LC (50 * 2 mm), and the column temperature is 30 ℃. Mobile phase A is 10 mM Ammonium Formate, and mobile phase B is Acetonitrile. LC-MS (Thermoscientific Q EXACTIVE PLUS) was used to analyze intermediates of PFOA qualitatively. The chromatographic column model used in LC-MS is Acquity Uplc @HSS T3 1.8 um (2.1×100 mm), and the column temperature is 25 ℃. Mobile phase C is 0.1% Ammonium Formate in H_2O, and mobile phase D is Acetonitrile. The chloride ion selective electrode was used to detect Cl^-. The standard solution was 0.1 $mol \cdot L^{-1}$ KCl, and the ion strength regulator was 1.0 $mol \cdot L^{-1}$ KNO_3. Sample testing used 1.0 mL of sample, 9.0 mL of ultrapure water, and 0.2 mL of ion strength regulator. The surface morphology was analyzed by SEM (SU8010, Hitachi, Japan). XPS (EscaLab 250Xi, Thermo Fisher Scientific, USA) was used to detect the valence state of the element in the interface of Fe_xS_y-coated quartz sand. The XPS data was calibrated by referencing the C 1s line to 284.8 eV. XPS Peak 41 application software was used to analyze XPS raw data. The crystallinity of Fe_xS_y and secondary mineral species were analyzed using powder X-ray diffraction (XRD, D8 Advance diffractometer, Bruker, Germany) with Cu Kα radiation (λ = 0.15418 nm).

(5) Model of PFOA cross-media transport and fate in microbe-iron sulfur minerals (Fe_xS_y) interaction media

The analysis of the breakthrough curves of PFOA in three types of interaction media was performed using HYDRUS-1D software. The two-site sorption coupling biological-chemical (microbe-Fe_xS_y) multi-process interaction model was constructed to simulate PFOA transport and fate behavior. The initial water flow parameters were provided in Table 5.9.

Table 5.9 The water flow parameters, which were obtained by the neural network prediction in the HYDRUS-1D.

Pedotransfer Function	Parameter
Input	
Sand (%)	100
Silt (%)	0
Clay (%)	0
Bulk Density ($g \cdot cm^{-3}$)	1.65
Output	
Q_r(-)	0.042
Q_s(-)	0.410
Alpha (cm^{-1})	0.145
n (-)	2.880
K_s($cm \cdot min^{-1}$)	0.660 *
l (-)	0.500

* The predicted saturation permeability coefficient *Ks* was 0.660 $cm \cdot min^{-1}$; the *Ks* was modified to 0.365, 0.417, 0.343, 0.380 $cm \cdot min^{-1}$ based on the PFOA breakthrough curves in Control, alone microbe (Cell&QS), alone Fe_xS_y(Fe_xS_y&QS), and microbe-Fe_xS_y(Cell&(Fe_xS_y&QS)) interaction media. $K = K_s \cdot Q_{s1}/Q_s$. Q_{s1} represented saturated water content, and it was 0.227, 0.259, 0.213, and 0.236, respectively, in Control, Cell&QS, Fe_xS_y&QS, and Cell&(Fe_xS_y&QS) interaction media. Q_{s1} is calculated by the formula: $Q_{s1} = (m_{wet} - m_{dry})/V_{column}$.
Q_r is residual soil water content;
Q_s is saturated soil water content;
Alpha is the parameter in the soil water retention function;
n is the parameter in the soil water retention function;
K_s is saturated hydraulic conductivity;
l is the tortuosity parameter in the conductivity function;
m_{wet}: wet weight of column with filler under saturated water condition;
m_{dry}: dry weight of column and filler without water;
V_{column}: volume of the column.

The revised standard equations can be written as (Wang et al., 2022):

$$\frac{\partial C}{\partial t} + f\frac{\rho}{\theta}K_d\frac{\partial C}{\partial t} + (1-f)\frac{\rho}{\theta}\frac{\partial S}{\partial t} = D\frac{\partial^2 C}{\partial z^2} - \nu\frac{\partial C}{\partial z} + \sum_k (k_{b_k} + \lambda_k)C, \tag{5-4}$$

$$\frac{\rho}{\theta}\frac{\partial S}{\partial t} = \alpha C. \tag{5-5}$$

Among them, C represents the concentration of PFOA, t represents time, f represents the proportion of instantaneous retardation to total retardation, ρ represents the dry bulk density ($M \cdot L^{-3}$), θ represents water content, K_d represents the instantaneous equilibrium adsorption coefficient ($L^3 \cdot M^{-1}$), S represents the concentration of PFOA in the solid phase ($M \cdot M^{-1}$), D represents the hydrodynamic dispersion coefficient ($L^2 \cdot T^{-1}$), z represents the vertical spatial coordinate (L), and v represents the velocity of pore water ($L \cdot T^{-1}$), λ represents the rate of chemical reaction (T^{-1}), k_b represents the biodegradation rate (T^{-1}), and α represents first-order sorption coefficient (T^{-1}).

5.4.3 Results and discussion

(1) Effect of microbe-Fe_xS_y interaction on cross-media transport and fate behavior of PFOA

Compared to alone Fe_xS_y (85.00%), alone microbe (86.00%), and Control (96.00%) media, only 67.14% of PFOA broke through in the microbe-Fe_xS_y interaction media (Figure 5.21 (a)), indicating that PFOA had the greatest attenuation rate in microbe-Fe_xS_y interaction media. Moreover, the sorption curves in 0 - 7 PVs stages showed that microbe-Fe_xS_y interaction media had the best sorption retardation effects for PFOA (Figure 5.21 (a)). Analyzing the change in pore volumes (PVs) (Figure 5.21 (a)), in comparison to the Control (0.227), the systems with microbe alone (Cell&QS) (0.259) and microbe-Fe_xS_y interaction (Cell&(Fe_xS_y&QS)) (0.236) experienced a 14.10% increase in porosity, whereas the Fe_xS_y&QS (0.213) media systems saw a 6.57% decrease (Table 5.9). This implied that initial-stage microbial growth loosens the media and makes it more porous, thereby elevating porosity (Ashad, 2023).

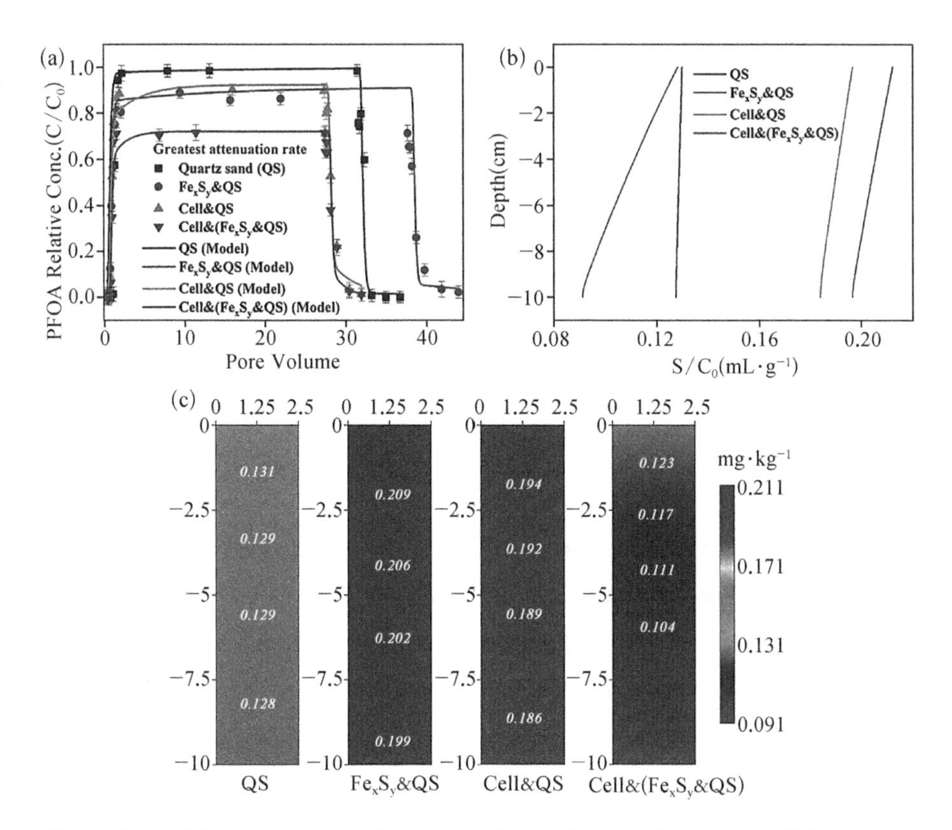

Figure 5.21 PFOA transport and fate behavior in Control (Quartz sand (QS)), alone microbe (Cell&QS), alone Fe_xS_y (Fe_xS_y&QS), and microbe-Fe_xS_y (Cell&(Fe_xS_y&QS)) interaction media (a); retained profiles (b) and 2D-spatial distribution (c) of PFOA in solid media.

Figure 5.21 (a) proved a great fit (more than 90%). It demonstrated that multivariate coupling of tracer and PFOA can obtain an accurate PFOA transport and transformation model. Upon model calibration, the model concurrently obtained the retention profiles (S/C_0, $mL\cdot g^{-1}$) (Figure 5.21 (b)) and solid phase distribution concentration (S, $mg\cdot kg^{-1}$) for PFOA (Figure 5.21 (c)). Figure 5.21 (b) shows the retained profiles of PFOA in solid media. The S/C_0 (the ratio of corresponding depth solid phase distribution concentration (S) to initial concentration (C_0)) of PFOA was changed and influenced by depth. In comparison to Control (QS) (0.128 $mL\cdot g^{-1}$–0.130 $mL\cdot g^{-1}$), alone microbe (Cell& QS) (0.184 $mL\cdot g^{-1}$–0.197 $mL\cdot g^{-1}$), and alone Fe_xS_y (Fe_xS_y& QS)

($0.197\ mL \cdot g^{-1}$–$0.212\ mL \cdot g^{-1}$), the S/C_0 ratio of PFOA within microbe-Fe_xS_y interaction media (Cell&(Fe_xS_y&QS)) ($0.0912\ mL \cdot g^{-1}$ – $0.128\ mL \cdot g^{-1}$) showed more pronounced depth-dependent variability.

Model results demonstrated that 2D-spatial distribution of PFOA in alone Fe_xS_y (Fe_xS_y&QS) ($0.212\ mg \cdot kg^{-1}$) and alone microbe (Cell&QS) ($0.197\ mg \cdot kg^{-1}$) media had a greater maximum retention or residual amount for PFOA, compared to microbe-Fe_xS_y(Cell&(Fe_xS_y&QS)) ($0.128\ mg \cdot kg^{-1}$) media. It revealed that microbe-Fe_xS_y interaction led to a decrease of PFOA retention or residual amount (Figure 5.21(c)). The phenomenon was consistent with experiment results (Figure 5.21(a)), in which microbe-Fe_xS_y interaction had the greatest PFOA attenuation rate.

The permeability coefficients *Ks* (Table 5.9) and dispersion *a* (Tables 5.10 and 5.11) were validated based on the breakthrough curves of PFOA (Figure 5.21(a)) and Cl^-(Figure 5.22). The model parameters for PFOA (Table 5.10) revealed the following potential mechanism: within the microbe-Fe_xS_y (Cell&(Fe_xS_y& QS)) interaction media, a notable enhancement in the attenuation rate λ ($0.083\ h^{-1}$ to $0.343\ h^{-1}$) and distribution coefficient K_d($0.349\ cm^3 \cdot g^{-1}$ to $0.391\ cm^3 \cdot g^{-1}$) was observed. Moreover, the fraction *f* denoting instantaneous equilibrium sorption

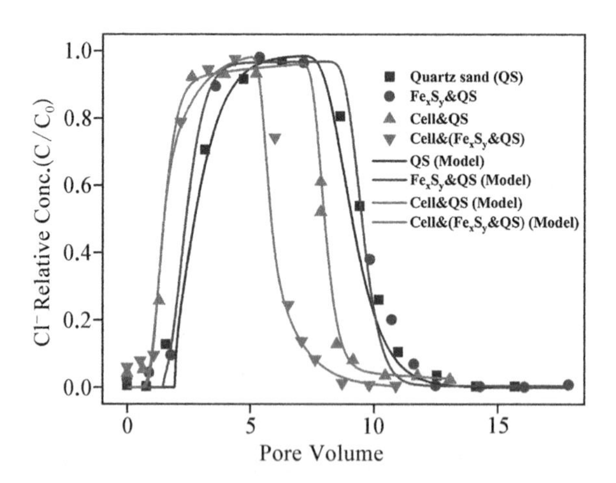

Figure 5.22 The breakthrough curves of KCl tracer in alone microbe (Cell&QS), alone Fe_xS_y(Fe_xS_y&QS), and microbe-Fe_xS_y (Cell&(Fe_xS_y&QS)) interaction media systems.

escalated from 28.8% to 47.9%, while the first-order kinetic sorption coefficient α surged from 0.039 to 0.160 h^{-1}(Table 5.10). It demonstrated that the augmentation of PFOA cross-media transport from the water phase to the solid phase in the microbe-Fe_xS_y interaction system was contributed by an increase in the proportion of instantaneous equilibrium adsorption and the obvious increase of first-order kinetic sorption rate. These combined outcomes substantiated that microbe-Fe_xS_y interaction media manifest the greatest attenuation rate and most effective cross-media adsorption retardation effect for PFOA. The relevant mechanism was discussed in the following Section (5) in this study.

Table 5.10 Parameters for PFOA in Control (quartz sand, QS), alone microbe (Cell&QS), alone Fe_xS_y(Fe_xS_y&QS), and microbe-Fe_xS_y(Cell&(Fe_xS_y&QS)) interaction media systems.

Variable		Disp (a/cm)	FRAC (f/−)	KD (K_d/$cm^3 \cdot g^{-1}$)	ALPHA (α/h^{-1})	SNKL1 (K_b or λ/h^{-1})	R^2
Interaction media type	Control (QS)	0.667	42%	0.241	0.023	—	91.16%
	Fe_xS_y&QS	0.327	29%	0.349	0.039	0.091	94.73%
	Cell&QS	0.478	45%	0.382	0.117	0.083	94.56%
	Cell&(Fe_xS_y&QS)	0.254	48%	0.391	0.160	0.343	95.35%
Combined ions-PFOA type	Cl^--Na^+-PFOA	0.336	29%	0.292	0.212	0.271	96.91%
	HCO_3^--PFOA	0.325	91%	0.294	0.218	0.159	97.00%
	NO_3^--PFOA	0.242	8%	0.447	0.583	0.252	95.79%
	SO_4^{2-}-PFOA	0.299	51%	0.437	0.376	0.226	98.89%

Table 5.11 Parameters for Cl^- in alone microbe (Cell&QS), alone Fe_xS_y (Fe_xS_y&QS), and microbe-Fe_xS_y(Cell&(Fe_xS_y&QS)) interaction media systems.

Variable	Quartz sand (QS)	Fe_xS_y&QS	Cell&QS	Cell& (Fe_xS_y&QS)
Disp (a/cm)	0.667	0.327	0.478	0.254
FRAC (f/−)	23%	16%	15%	26%
KD (K_d/$cm^3 \cdot g^{-1}$)	0.109	0.189	0.077	0.110
ALPHA(α/h^{-1})	0.002	0.004	0.021	0.128
R^2	95.29%	89.04%	89.60%	89.10%

(2) Specific effect of PFOA-ions occurrence types on PFOA cross-media transport and fate in microbe-Fe_xS_y interaction media

Figure 5.23 (a) demonstrated that compared to PFOA-HCO_3^-(83.19%) and PFOA-SO_4^{2-}(78.44%) occurrence environment, less PFOA broke through the microbe-Fe_xS_y interaction media under the PFOA-Cl^--Na^+(72.90%) and PFOA-NO_3^-(73.50%) occurrence environment. Notably, across all four PFOA-ions occurrence environments, an inhibiting effect on the attenuation rate of PFOA within microbe-Fe_xS_y interaction media was evident when compared to the Control. Remarkably, the presence of HCO_3^- exerted the most potent inhibiting effect, resulting in a significant decrease in λ from 0.343 h^{-1} to 0.159 h^{-1}. Additionally, NO_3^- and SO_4^{2-} demonstrated pronounced adsorption-desorption retardation effects, notably in the initial 0 – 5 PVs stage and the subsequent 28 PVs –33 PVs stage.

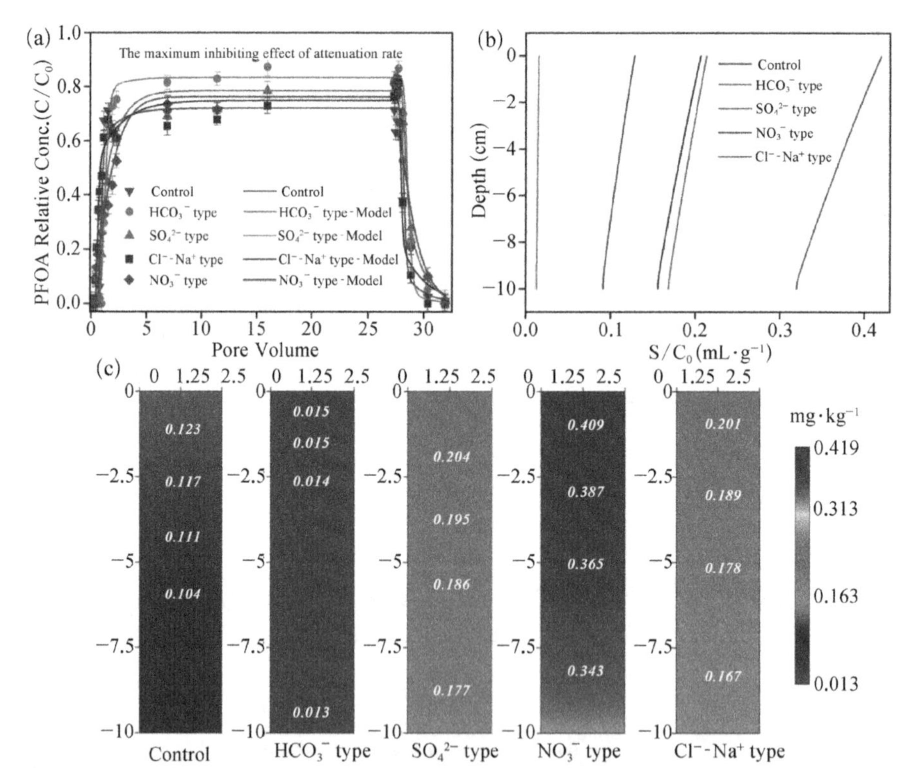

Figure 5.23 Effect of PFOA-ions occurrence type on PFOA transport and fate behavior in microbe-Fe_xS_y(Cell&(Fe_xS_y&QS)) interaction-induced media (a); retained profiles (b) and 2D-spatial distribution (c) of PFOA in solid media; Control was an ion free system with PFOA.

Figure 5.23 (b) distinctly indicated the dramatic differences in the spatial distribution of PFOA. Compared to PFOA-HCO_3^-(0.0130 mL·g^{-1}–0.0155 mL·g^{-1}) occurrence environment, the S/C_0 of PFOA in PFOA-SO_4^{2-}(0.169 mL·g^{-1}–0.214 mL·g^{-1}), PFOA-NO_3^-(0.320 mL·g^{-1}–0.420 mL·g^{-1}) and PFOA-Cl^--Na^+ (0.156 mL·g^{-1}–0.206 mL·g^{-1}) occurrence environments experienced more pronounced declines with increasing depth.

Figure 5.23 (c) showed that PFOA-NO_3^-(0.320 mg·kg^{-1}–0.420 mg·kg^{-1}) and PFOA-SO_4^{2-}(0.169 mg·kg^{-1}–0.214 mg·kg^{-1}) occurrence environments had greater adsorption performance of PFOA in solid media than that in PFOA-Cl^--Na^+ (0.156 mg·kg^{-1}–0.206 mg·kg^{-1}) and PFOA-HCO_3^- (0.0130 mg·kg^{-1}–0.0155 mg·kg^{-1}) occurrence environments. Model fitting results (Table 5.10) revealed that the K_d was relatively lower in PFOA-HCO_3^- (0.294 cm^3·g^{-1}) and PFOA-Cl^--Na^+ (0.292 cm^3·g^{-1}) occurrence environments as compared to PFOA-SO_4^{2-} (0.437 cm^3·g^{-1}) and PFOA-NO_3^- (0.447 cm^3·g^{-1}) occurrence environments. Additionally, the dispersity a decreased from 0.254 to 0.242 cm in the PFOA-NO_3^- occurrence environment. The first-order kinetic sorption coefficient α increased from 0.160 h^{-1} to 0.583 h^{-1} and 0.374 h^{-1} in PFOA-NO_3^- and PFOA-SO_4^{2-} occurrence environments, respectively (Table 5.10). It demonstrated that the augmentation of PFOA cross-media transport in PFOA-NO_3^- occurrence environments contributed by the obvious increase of instantaneous equilibrium adsorption and first-order kinetic sorption rate. The fraction f decreased in PFOA-NO_3^- occurrence environments, indicating an increase in the proportion of first-order kinetic sorption. Moreover, in PFOA-SO_4^{2-} occurrence environments, the fraction f was not influenced by the augmentation of PFOA cross-media transport contributed by the obvious increase of instantaneous equilibrium adsorption and first-order kinetic sorption rate. The relevant mechanism was discussed in following Section (5) in this study.

Overall, the existence of the above four ions can inhabit the attenuation rate of PFOA in microbe-Fe_xS_y interaction media, and the PFOA-HCO_3^- occurrence environment showed the greatest inhibiting effect. Moreover, PFOA-NO_3^- and PFOA-SO_4^{2-} occurrence environments showed enhanced adsorption-desorption retardation effects for PFOA in microbe-Fe_xS_y interaction media. This highlighted that PFOA exhibited the greatest potential transport risk within the microbe-Fe_xS_y interaction media under the PFOA-HCO_3^- occurrence environment.

(3) Secondary minerals transformation in microbe-Fe_xS_y interaction media

The iron-sulfur/oxide crystal structure of Fe_xS_y&QS was depicted as a stripe cluster in the SEM (Figure 5.24 (a)-(b)). SEM element mapping results indicated an even distribution of Fe and S elements across the surface of the quartz sand (Figure 5.24 (c)). Additionally, the SEM findings revealed that the secondary minerals appeared in other structures as the stripe cluster structure vanished after the reaction (Figure 5.24 (d)-(g)). After the reaction, the surface of the Cell&(Fe_xS_y&QS) media exhibited distinct changes: a smooth clavate-like structure emerged in the PFOA-HCO_3^- occurrence environment, rod-shaped aggregates

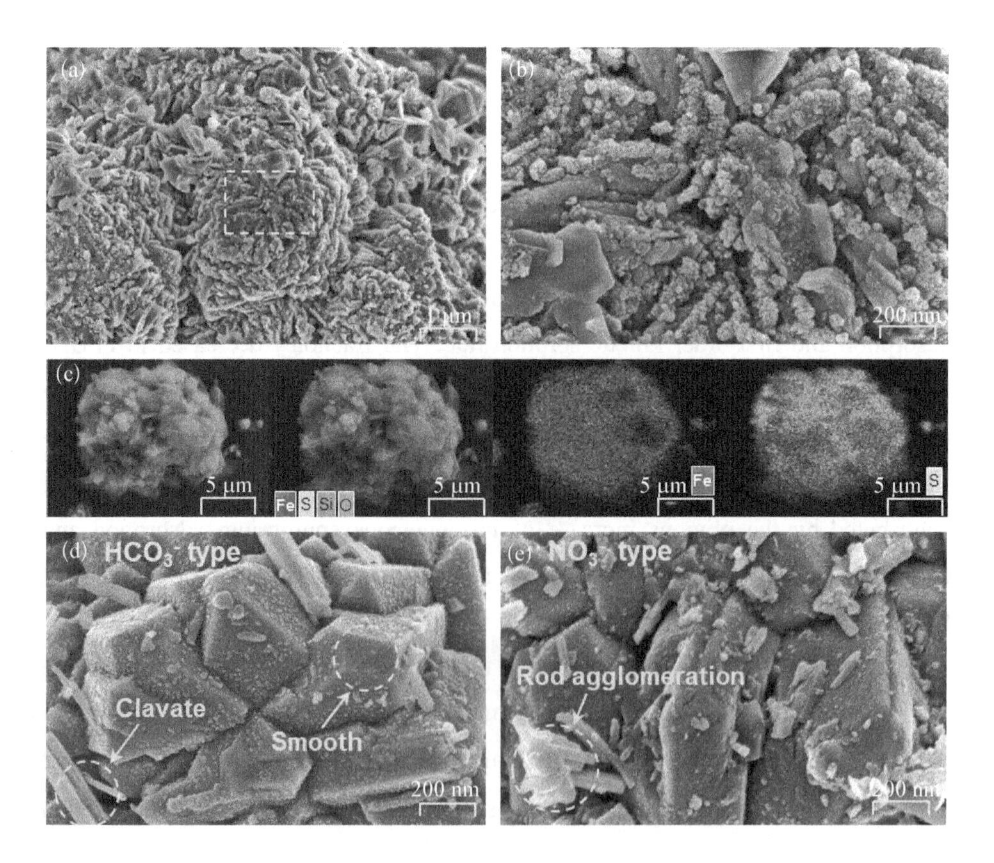

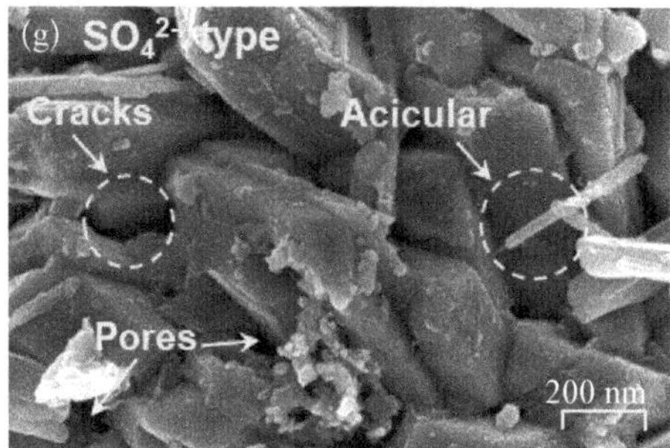

Figure 5.24 SEM images of Fe_xS_y-coated quartz sand before reaction (a)-(b); SEM-mapping of Fe_xS_y-coated quartz sand before reaction (c); SEM images of Fe_xS_y-coated quartz sand after reaction from microbe-Fe_xS_y(Cell&(Fe_xS_y&QS)) interaction media in PFOA-HCO_3^-, PFOA-NO_3^-, PFOA-Cl^--Na^+, PFOA-SO_4^{2-} ions occurrence environments (d)-(g).

appeared in the PFOA-NO_3^- occurrence environment, rod-shaped structures with accompanying pores were seen in the PFOA-Cl^--Na^+ occurrence environment and acicular-like structures with cracks and pores were observed in the PFOA-SO_4^{2-} occurrence environment (Figure 5.24 (d)-(g)).

The XPS spectra of Cell&(Fe_xS_y& QS) in the four types of PFOA-ions occurrence environments before and after the reaction were presented in Figure 5.25. Fe(II)-S and Fe(III)-S were credited with causing the 707.0 eV and 710.6 eV Fe 2p3/2 peak, as well 719.7 eV and 724.2 eV Fe 2p1/2 peak, respectively (Zazpe et al., 2023; Zhao et al., 2020). Fe(III)-O was responsible for the Fe 2p3/2 peak at 713.1 eV and Fe 2p1/2 peak 727.0 eV (Chubar et al., 2021; Zhao et al., 2020). Fe(II)-S (51.81%), Fe(III)-S (29.26%), and Fe(III)-O (18.93%) were detected in the range of Fe 2p before the reaction (Figure 5.25 (a)).

Following the reaction, the content of Fe(II)-S on the Cell&(Fe_xS_y&QS) interface decreased, especially, in PFOA-NO_3^- occurrence environments, the content of Fe(II)-S decreased dramatically (51.81% to 19.03%). In addition, the content of Fe(III)-S (29.26% to 51.90%) and Fe(III)-O (18.93% to 41.73%) increased, which suggested that Fe^{2+} from Fe(II)-S minerals was oxidized into Fe^{3+}, then subsequently formed Fe(III)-S/Fe(III)-O minerals or Fe(III)-SO_4^{2-} after the reaction in different PFOA-ions occurrence environments (Figure 5.25(c),

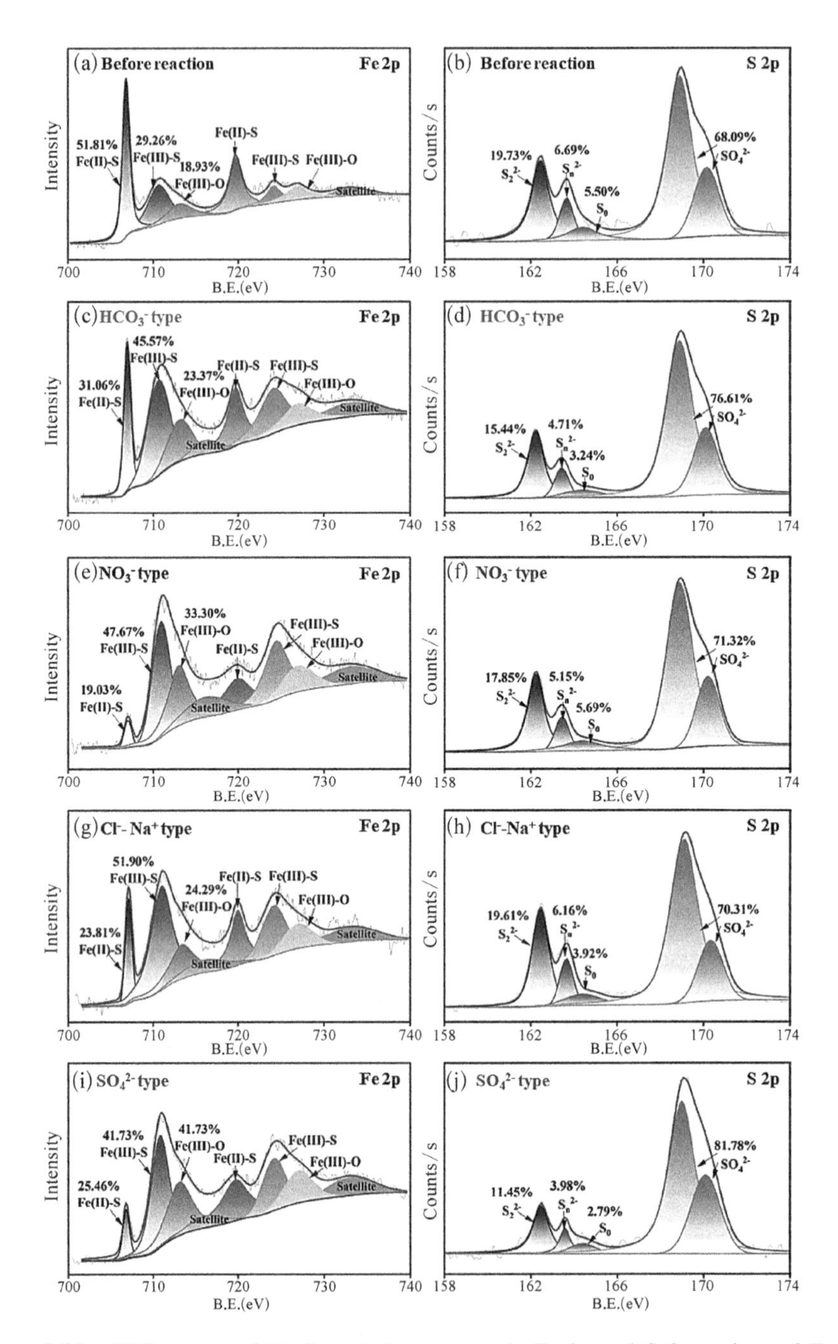

Figure 5.25 XPS spectra of Fe_xS_y-coated quartz sand: Fe 2p and S 2p regions of Fe_xS_y-coated quartz sand before reaction (a)–(b); Fe 2p and S 2p regions of Fe_xS_y-coated quartz sand after reaction from microbe- Fe_xS_y(Cell&(Fe_xS_y&QS)) interaction system in PFOA-HCO_3^-, PFOA-NO_3^-, PFOA-Cl^--Na^+, PFOA-SO_4^{2-} ions occurrence environments (c)–(j).

5.25(e), 5.25(g), 5.25(i)). In addition, S_2^{2-}(19.73%), S_n^{2-}(6.69%) and S_0(5.50%) were responsible for the peaks of S 2p at 162.4 eV, 163.6 eV and 164.4 eV, respectively, while SO_4^{2-}(68.09%) was responsible for the peaks of S 2p at 168.9 eV and 170.2 eV before reaction (Figure 5.25 (b)) (Zhao et al., 2020). After the reaction, the S_2^{2-} content on the surface of Cell&(Fe_xS_y&QS) decreased in four PFOA-ions occurrence environments (Figure 5.25(d), 5.25(f), 5.25(h), 5.25(j)). The content of SO_4^{2-} increased obviously (68.09% to 76.61%, 81.78%) in PFOA-HCO_3^- and PFOA-SO_4^{2-} occurrence environments than that in PFOA-Cl^--Na^+ and PFOA-NO_3^- occurrence environment (Figure 5.25 (h)). However, the content of Fe(III)-S increased drastically from 18.93% to 51.90% and 47.67% (Figure 5.25 (g)) in the PFOA-Cl^--Na^+ and PFOA-NO_3^- occurrence environment. It suggested that the increase of Fe(III)-S content can be attributed to the formation of other minerals (e.g. $Fe_{n-1}S_n$, $Fe_{1-n}S$ minerals (XRD confirmed this hypothesis)).

According to the XRD results (Figure 5.26), the surface composition of Cell&(Fe_xS_y&QS) before the reaction primarily was iron-sulfur minerals FeS_2 (JCPDS#2-908), $Fe_{1-n}S$ (JCPDS#22-1120), $Fe_2(SO_4)_3$(JCPDS#42-225) and iron oxide mineral β-$Fe_2O_3 \cdot H_2O$ (JCPDS#3-440). This observation aligned well with the findings from the XPS analysis (Figure 5.25 (a)-(b)).

After the reaction, β-$Fe_2O_3 \cdot H_2O$ was converted to α-Fe_2O_3(JCPDS#5-637) in the PFOA-HCO_3^- occurrence environment. In the PFOA-Cl^--Na^+ occurrence environment, the transformation led to the formation of γ-$Fe_2O_3 \cdot H_2O$ (JCPDS# 2-127) and $Fe_{n-1}S_n$(JCPDS#3-1028). The PFOA-NO_3^- occurrence environment prompted the generation of α-$Fe_2O_3 \cdot H_2O$ (JCPDS#8-97) and Fe_9S_{10}(JCPDS#35-1043). In the PFOA-SO_4^{2-} occurrence environment, β-$Fe_2O_3 \cdot H_2O$ was the main mineral (Figure 5.26). However, compared to PFOA-HCO_3^- (23.37%) and PFOA-Cl^--Na^+(24.29%) occurrence environment, the content of Fe(Ⅲ)-O in PFOA-SO_4^{2-} (41.73%) and PFOA-NO_3^- (33.30%) occurrence environments significantly increased. It indicated that more iron oxide minerals were generated in PFOA-SO_4^{2-} and PFOA-NO_3^- occurrence environments. The XRD results revealed that secondary minerals were more inclined to transform into α-$Fe_2O_3 \cdot H_2O$ in the

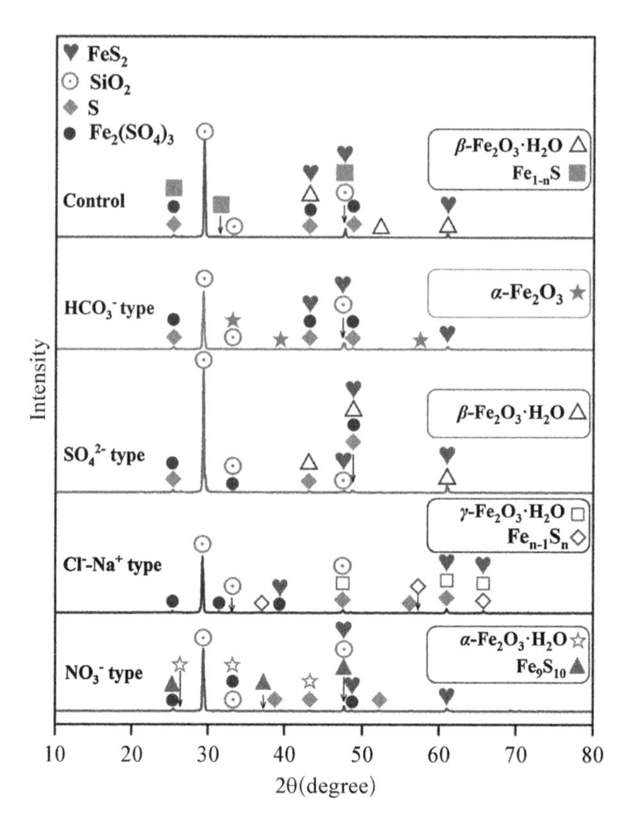

Figure 5. 26 XRD patterns of the Fe_xS_y-coated quartz sand before (Control) and after the reaction from microbe-Fe_xS_y(Cell&(Fe_xS_y&QS)) interaction system in PFOA-HCO_3^-, PFOA-NO_3^-, PFOA-Cl^--Na^+, PFOA-SO_4^{2-} ions occurrence environments.

PFOA-NO_3^- occurrence environment, and β-$Fe_2O_3 \cdot H_2O$ in the PFOA-SO_4^{2-} occurrence environment. These transformations in the mineral structure were likely influenced by the specific PFOA-ions occurrence environments and their interactions with the microbe-Fe_xS_y system. The relevant mechanism was discussed in following Section (5) in this study.

(4) PFOA attenuation pathways in microbe-Fe_xS_yinteraction media

Based on LC-MS findings and prior studies, it was speculated that the potential attenuation pathway for PFOA (Figure 5.27) was: the chain reaction (I) of Deprotonation (A) with the cycle of Activation (B), Decarboxylation (C), Hydroxylation (D), HF elimination (E), Hydrolysis (F), and HF elimination

(E) ; the cycle chain reaction (II) of Decarboxylase (G) , Hydroxylase (H) , HF elimination (E) , and Hydrolase (I) ; and the chain reaction (III) of Deprotonation (A) , Activation (B) , F^- elimination (K-K) twice and C-C bond scission (J) (Ma et al., 2017; Xiao et al., 2023; Yang et al., 2022). In the above pathways, the Activation step (B) was mainly achieved by Fe^{2+} and $S_2{}^{2-}$ as electron donors to promote the formation of perfluorooctanoic acid free radical (F-$(CF_2)_7$-COO ·) from F-$(CF_2)_7$-COO^-.

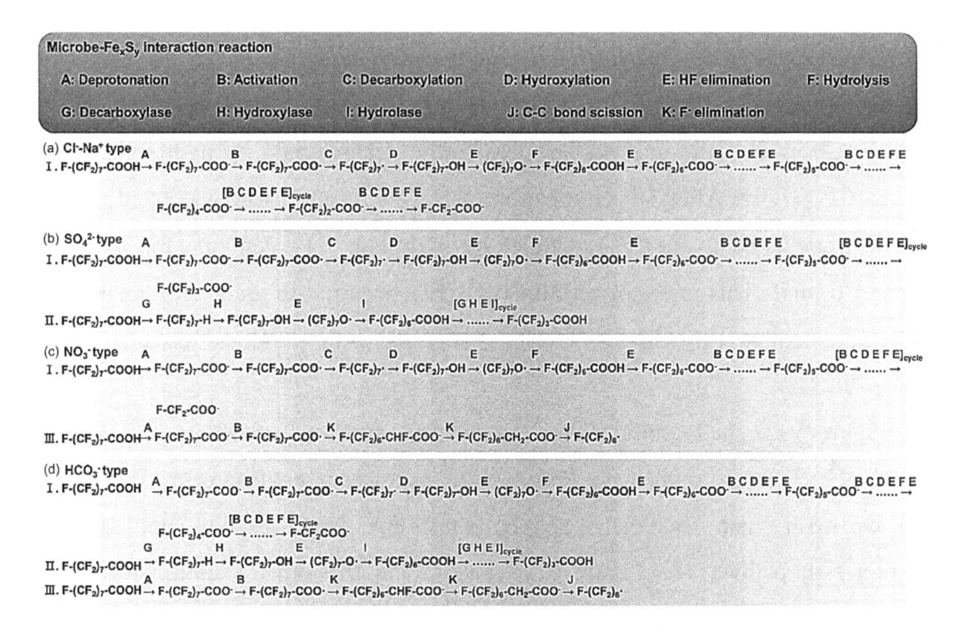

Figure 5.27 Potential fate pathway of PFOA considering microbe-Fe_xS_y interaction in PFOA-Cl^--Na^+(Yang et al., 2022) , PFOA-$SO_4{}^{2-}$(Xiao et al., 2023; Yang et al., 2022) , PFOA-$NO_3{}^-$(Ma et al., 2017; Yang et al., 2022) and PFOA-$HCO_3{}^-$(Ma et al., 2017; Xiao et al., 2023; Yang et al., 2022) ions occurrence environments. The intermediates marked in red were detected in the experiments.

Figure 5. 27 (a) illustrated the attenuation pathway of PFOA considering microbe-Fe_xS_y interaction in the PFOA-Cl^--Na^+ occurrence environment. PFOA (F-$(CF_2)_7$-COOH) was first converted to F-$(CF_2)_7$-COO^- by deprotonation (A) , and it was activated (B) to F-$(CF_2)_7$-COO · , which then generated F-$(CF_2)_7$ · and CO_2 by decarboxylation (C). Then hydroxylation (D) , HF elimination (E) , hydrolysis (F), and HF elimination (E) reactions followed, and the target product F-$(CF_2)_6$-COO^- was detected by LC-MS. In addition, a series of intermediates and the shortest carbon chain intermediate including F-CF_2-COO^- were also detected, which was consistent with the research results of Yang et al (Yang et al., 2022).

Figure 5.27 (b) was the attenuation pathway of PFOA considering microbe-Fe_xS_y interaction in the PFOA-SO_4^{2-} occurrence environment. This pathway comprised two primary cycle chain reactions, denoted I and II. The former represented cycle chain reaction I in pathway a (Figure 5.27 (a)). The latter initiated with PFOA (F-$(CF_2)_7$-COOH) being catalyzed by decarboxylase (G) to yield F-$(CF_2)_7$-H. This intermediate was further transformed by hydroxylase (H) into F-$(CF_2)_7$-OH. After HF elimination (E), $(CF_2)_7$-O · was generated, which was then converted to F-$(CF_2)_6$-COOH by hydrolase (I), denoting the reduction of one CF_2 unit. This series of reactions (GHEI) occurred to yield additional short-chain intermediate (F-$(CF_2)_3$-COOH), aligning with previous research (Xiao et al., 2023).

Figure 5.27 (c) outlined the attenuation pathway of PFOA considering microbe-Fe_xS_y interaction in the PFOA-NO_3^- occurrence environment. There were also two main chain reactions I and III for PFOA. One was the same as chain reaction I in pathway a (Figure 5.27 (a)), and the other started with PFOA (F-$(CF_2)_7$-COOH) undergoing deprotonation (A) and activation (B), generating F-$(CF_2)_7$-COO · free radical. In succession, F-$(CF_2)_7$-COO · was subject to successive hydrated electron-induced F^- eliminations (KK), culminating in the formation of F-$(CF_2)_6$-CH_2-COO^-. The ensuing C-C bond scission (J) generated F-$(CF_2)_6$ · (Ma et al., 2017).

Figure 5.27 (d) was the attenuation pathway of PFOA considering microbe-Fe_xS_y interaction in the PFOA-HCO_3^- occurrence environment. There were three chain reactions for PFOA, which were the same as cycle chain reaction I in pathway a (Figure 5.27 (a)), chain reaction II in pathway b (Figure 5.27 (b)), and chain reaction III in pathway c (Figure 5.27 (c)) (Ma et al., 2017; Xiao et al., 2023; Yang et al., 2022).

(5) Mechanism of PFOA attenuation and secondary minerals transformation

Based on the results of Figures 5.21 – 5.27 and previous reports about the fate pathway of PFOA (Ma et al., 2017; Xiao et al., 2023; Yang et al., 2022), mineral transformation (Hellsing et al., 2016; Lyu et al., 2019), microbe and mineral/iron ions interaction (Percak-Dennett et al., 2017; Xu et al., 2020), the mechanisms of PFOA cross-media transport and fate and minerals transformation considering the microbe-Fe_xS_y interaction in four PFOA-ions interaction environments were proposed (Figure 5.28).

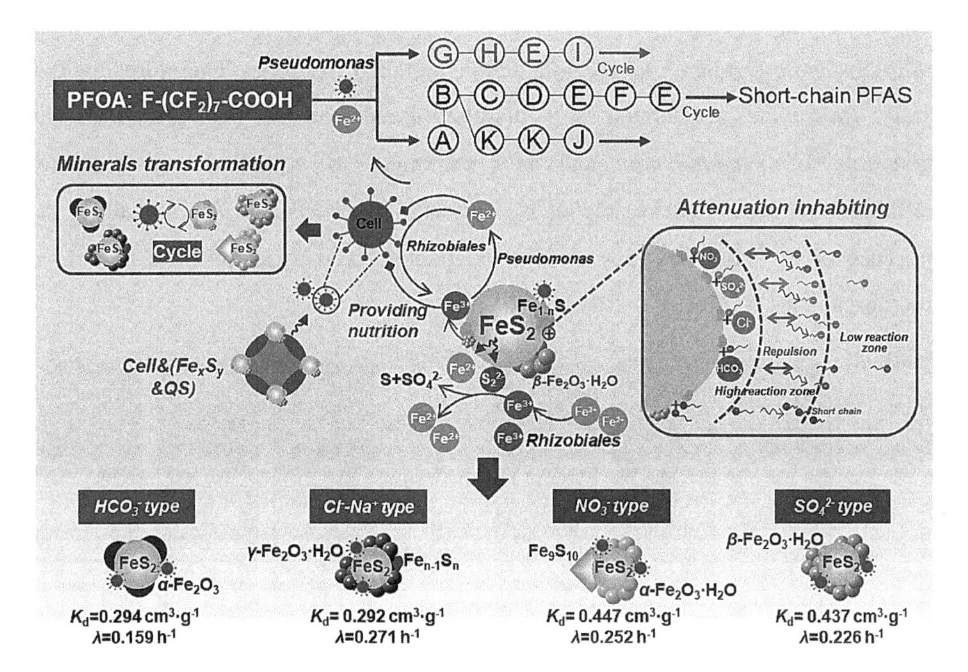

Figure 5.28 The proposed mechanism of PFOA transport and fate (Campos-Pereira et al., 2020; Wallace et al., 2019) and minerals transformation (Hellsing et al., 2016; Lyu et al., 2019) considering the microbe-Fe_xS_y interaction (Percak-Dennett et al., 2017) in four PFOA-ions occurrence environments.

Figure 5.21 showed that the PFOA attenuation rate λ increased by 277% in microbe-Fe_xS_y interaction media (0.343 h^{-1}) than that in alone Fe_xS_y media (0.091 h^{-1}). It probably resulted from the dual role of microbial interaction with minerals. *Pseudomonas* with an abundance of 1.63% was detected in this study (Figure 5.29). It was reported that the intracellular metabolism of *Pseudomonas* can facilitate the transformation of Fe^{3+} to Fe^{2+} (Xu et al., 2020). *Pseudomonas* has also been documented to have the ability to transform fluoromodulated alcohols into less harmful short-chain perfluorocarboxylic acids by eliminating the -CF_2- moiety stepwise (Kim et al., 2012). Another functional microbe *Rhizobiales* with an abundance of 0.5% was also identified in our study (Figure 5.29). *Rhizobiales* can catalyze the enzyme-driven oxidation of Fe^{2+} to Fe^{3+}, and react with the S_2^{2-} site of FeS_2 to attack (dissolve) pyrite, resulting in the release of SO_4^{2-} and S, accompanied by the regeneration of Fe^{2+} (Percak-Dennett et al., 2017). Figure 5.28 showed that two Fe^{2+} ions were consumed by *Rhizobiales* and three Fe^{2+} ions were produced in the interaction system each time. Therefore, in this study, *Rhizobiales* accelerated Fe_xS_y dissolution and more Fe^{2+} was produced. The dual role of *Pseudomonas* and Fe^{2+} combined to complete fast reductive defluorination and remove the -CF_2- moiety from PFOA. Additionally, the presence of Fe^{2+} and Fe^{3+} ions as trace elements served to support the growth of microorganisms (Ng et al., 2016).

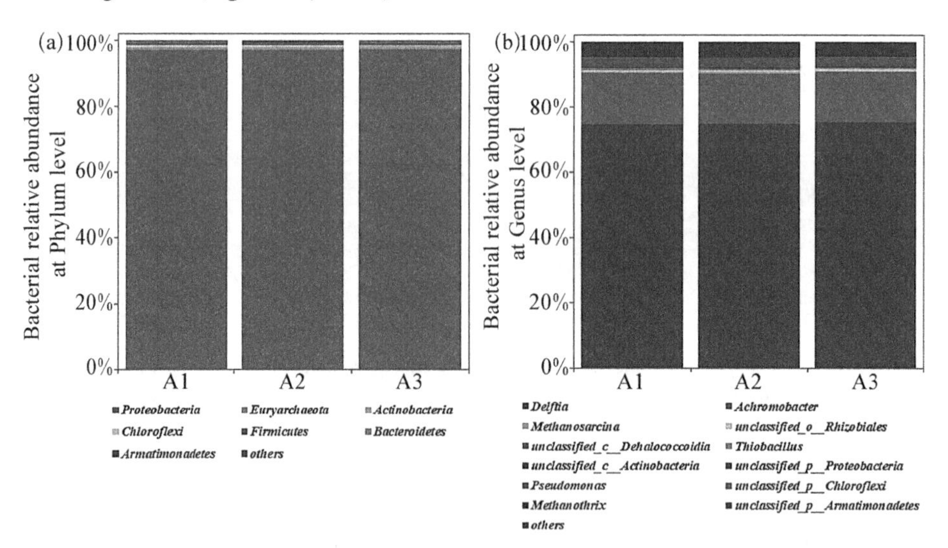

Figure 5.29 Diversity and abundance of bacteria at the phylum (a) and the genus (b) levels under PFOA exposure.

Moreover, LC-MS results showed that the three predominant attenuation pathways of PFOA encompassed: the chain reaction (I) of A with the cycle of B, C, D, E, F, and E (Yang et al., 2022); the cycle chain reaction (II) of G, H,

E, and I (Xiao et al., 2023); and the chain reaction (III) of A, B, K, and J (Ma et al., 2017). *Pseudomonas* was involved in the cycle chain reaction (II), in which G (Decarboxylase), H (Hydroxylase), and I (Hydrolase) were the main involved processes (Xiao et al., 2023). Collaboratively, the microbe-Fe_xS_y interaction drove multi-path chain reaction decay cycles PFOA and controlled secondary minerals-ions (Fe^{2+}/Fe^{3+}) transformation cycles.

Figure 5.21 showed that K_d and α increased by 13.03% and 310.26% in microbe-Fe_xS_y media than that in alone Fe_xS_y media, respectively. XRD results (Figure 5.30) showed that after the reaction, β-$Fe_2O_3 \cdot H_2O$ and $Fe_{1-n}S$ transformed, leading to the formation of ε-Fe_2O_3 in alone Fe_xS_y media. Besides,

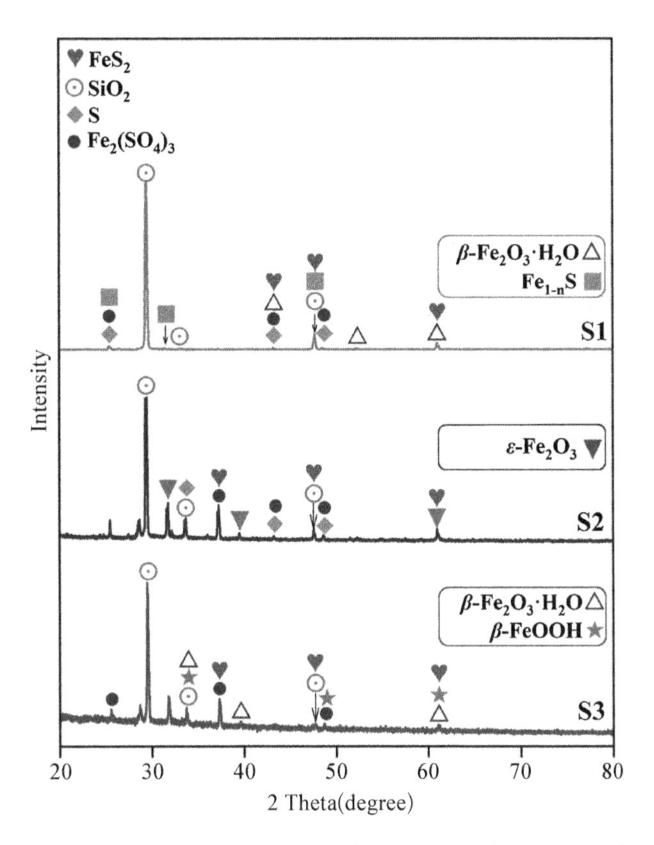

Figure 5.30 XRD patterns of Fe_xS_y-coated quartz sand before and after the reaction; S1, S2, and S3 represented Fe_xS_y&QS medium before reaction, Fe_xS_y&QS, and Cell&(Fe_xS_y&QS) system after reaction, respectively.

$Fe_{1-n}S$ disappeared, while β-$Fe_2O_3 \cdot H_2O$ persisted in microbe-Fe_xS_y media. This difference can likely be attributed to the enhanced adsorption and retardation effects exerted by β-$Fe_2O_3 \cdot H_2O$ and β-FeOOH (Campos-Pereira et al., 2020). Moreover, Figure 5.23 showed that SO_4^{2-}/NO_3^--microbe-Fe_xS_y interaction had greater retardation effects on PFOA. Based on the results of XPS (Figure 5.25) and XRD (Figure 5.26), upon completion of the reactions, a portion of FeS_2 and $Fe_{1-n}S$ transformed, resulting in more generation of α-$Fe_2O_3 \cdot H_2O$ in the PFOA-NO_3^- occurrence environment and β-$Fe_2O_3 \cdot H_2O$ in the PFOA-SO_4^{2-} occurrence environment (Figures 5.25 and 5.26). α-$Fe_2O_3 \cdot H_2O$ and β-$Fe_2O_3 \cdot H_2O$ led to K_d of PFOA increased from 0.292 $cm^3 \cdot g^{-1}$ to 0.437 $cm^3 \cdot g^{-1}$, 0.447 $cm^3 \cdot g^{-1}$ (Table 5.10). Previous research has indicated that compared to α-Fe_2O_3(0.10 $mg \cdot g^{-1}$), ferrihydrite ($Fe_2O_3 \cdot nH_2O$) enhanced adsorption potential (0.99 $mg \cdot g^{-1}$) for fluoride (Wallace et al., 2019). The unique characteristics of ferrihydrite, including its huge surface area and intricate pore structure, enable it to provide a broad surface and abundant adsorption sites available for PFOA adsorption (Xu et al., 2019).

Furthermore, it has been reported that anions competed for adsorption sites with PFOA (Wang et al., 2012). This competition resulted in the reduction of contact between PFOA and Fe^{2+} originating from $FeS_2/Fe_{1-n}S$. Consequently, fewer PFOA were adsorbed onto the mineral surfaces, primarily due to the presence of NO_3^-, SO_4^{2-}, HCO_3^-, and Cl^- ions occupying adsorption sites within the high reaction zone (Figure 5.28). Especially, HCO_3^- occupied those sites, leading to a greater electrostatic repulsion effect on PFOA. This resulted in the distribution of PFOA over a greater distance and in areas with lower concentrations of Fe^{2+} and microorganisms (low reaction zone) than in other anion environments (Figure 5.28). It showed a maximum inhibiting effect on PFOA attenuation (λ decreased from 0.343 h^{-1} to 0.159 h^{-1}) (Figure 5.23). This phenomenon contributed to the inhibiting effect on PFOA attenuation under the PFOA-ions occurrence environment. Therefore, the attenuation rate of PFOA in all occurrence environments decreased.

(6) Conclusions of this case study

The findings unveiled that the microbe-Fe_xS_y (pyrite and pyrrhotite) interaction media exhibited the greatest attenuation rate and the best adsorption retardation effect on PFOA. However, the presence of four types of ions inhabited the attenuation rate of PFOA, with the PFOA-HCO_3^- occurrence environment exhibiting the most pronounced inhibiting impact. NO_3^-, SO_4^{2-}, HCO_3^-, and Cl^- ions occupied adsorption sites and electrostatically repelled PFOA, resulting in the PFOA being repelled (electrostatic repulsion) and distributed in Fe^{2+}/microorganism-poor area (low reaction zone). Moreover, it is noteworthy that the transformation of iron-sulfur/oxide minerals exhibited a significant dependency on the types of PFOA-ion occurrences. In the presence of PFOA-NO_3^-, there was a notable increase in the emergence of α-$Fe_2O_3 \cdot H_2O$ and Fe_9S_{10}. Conversely, in the PFOA-SO_4^{2-} occurrence environment, a higher prevalence of β-$Fe_2O_3 \cdot H_2O$ was observed. In addition, when SO_4^{2-} or NO_3^- interacted with microbe-Fe_xS_y, they exhibited more pronounced retardation effects on PFOA. Furthermore, considering the microbe-mineral interaction, the attenuation pathways of PFOA showed diverse chain reaction cycles in different PFOA-ions occurrence environments. The maximum number of three-cycle attenuation pathways of PFOA was identified in the PFOA-HCO_3^- occurrence environment. Functional microbe *Pseudomonas* displayed a dual role in the reduction of Fe^{3+} to Fe^{2+} and -CF_2-moiety removal of PFOA. Besides, another functional microbe *Rhizobiales* contributed to the production of three Fe^{2+} after consuming two Fe^{2+} in microbe-Fe_xS_y interaction. Collaboratively, Fe^{2+} and *Pseudomonas* combined to drive multi-path chain reaction decay cycles of PFOA. This study provided a theoretical basis for understanding the cross-media transport and fate of PFOA with microbe-mineral interaction in various ion occurrence environments (Figure 5.31).

5.5 Conclusions and suggestions

Microorganisms accelerate the e^- transfer of electron donor groups; other non-metallic fillers are conducive to maintaining and improving the diversity and abundance of microorganisms. The above researches have demonstrated that microbe-mineral interaction Bio-PRBs are effective, however, some matters need

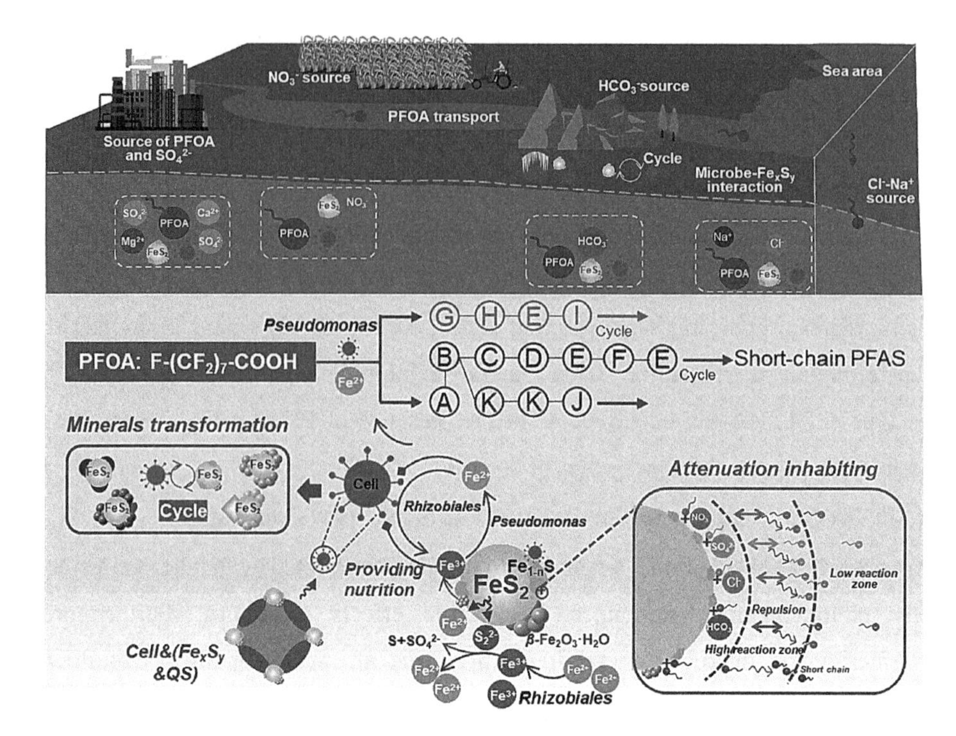

Figure 5.31 Cross-media transport and fate of PFOA with microbe-mineral interaction in various ion occurrence environments.

attention. Suggestions on the application and development of Bio-PRBs are proposed.

a. Considering the complexity of the actual site and the coexistence of combined pollutants (heavy metals, chlorinated hydrocarbons, and emerging contaminants) in groundwater and soil, portable field-scale combined pollutants remediation devices and systems for pilot-scale or field-scale application are urgently needed.

b. Surface runoff in the chemical plant area always contains large amounts of emerging contaminants, which can infiltrate and significantly contaminate the soil and groundwater. Constructing and designing novel pilot-scale or field-scale vertical reaction columns with drainage wells to alleviate emerging contaminants of surface runoff contamination are proposed.

Chapter 6
Multi-process Coupling Model for Microbe-mineral Interaction

6.1 Physical, chemical, and biological multi-process coupling model

Deeply understanding and accurately predicting pollution plume evolution with time changes in different scales of environmental media is particularly critical. Contaminant transport and fate behavior can be investigated and simulated through models and numerical simulation. Physical, chemical, and biological multi-process coupling reactions include convection, dispersion, adsorption, chemical reaction, and (or) biodegradation processes.

The convection, dispersion, and adsorption coupling models include two-site model (TSM) (Xing et al., 2021), continuous-distribution multi-rate (CDMR) model (Brusseau et al., 2019), modified-TSM (Silva et al., 2020), multi-process mass-transfer (MPMT) model (Brusseau, 2020), tempered one-sided stable density transport (TOSD) model (Zhou et al., 2021) and comprehensive compartment model (CCM) (Brusseau et al., 2019). The above model coupled with chemical reaction and (or) biodegradation modules to construct physical, chemical, and biological multi-process coupling models. Model governing equations were provided as follows.

(1) TSM coupling with chemical reaction and biodegradation

The two-site model (TSM) can be used to describe chemical nonequilibrium actions. Chemical nonequilibrium action is caused by the dynamic exchange adsorption and delayed adsorption of ions. The two-site adsorption theory hypothesizes that the adsorption sites in the media can be segmented into two types: the first type of site (type-1 site) is an instantaneous equilibrium adsorption site, and the remaining fraction of site (type-2 site) is the rate-constrained nonequilibrium adsorption site (Kamra et al., 2001). The adsorption process is assumed to be time-dependent and follows first-order kinetic equations (Genuchten, 1981; Van Genuchten and Wagenet, 1989).

When the multi-process model considers linear adsorption (K_d), single kinetic adsorption (α), biodegradation, and chemical reaction under microbe-mineral interaction conditions, the revised standard equations of convection, dispersion, adsorption, chemical reaction, and biodegradation can be written as (Wang et al., 2022)

$$\frac{\partial C}{\partial t} + f\frac{\rho}{\theta}\frac{\partial S_1}{\partial t} + (1-f)\frac{\rho}{\theta}\frac{\partial S_2}{\partial t} = D\frac{\partial^2 C}{\partial z^2} - v\frac{\partial C}{\partial z} + \sum_k (k_{b_k} + \lambda_k)C, \quad (6-1)$$

$$\frac{\partial S_1}{\partial t} = K_d\frac{\partial C}{\partial t}, \quad (6-2)$$

$$\frac{\rho}{\theta}\frac{\partial S_2}{\partial t} = \alpha C. \quad (6-3)$$

Where 1 and 2 represent the type-1 and type-2 sites, respectively, C represents the solute concentration ($M \cdot L^{-3}$), f represents the proportion of instantaneous retardation to total retardation, θ represents water content, t represents time, ρ represents the dry bulk density ($M \cdot L^{-3}$), K_d represents the instantaneous equilibrium adsorption coefficient ($L^3 \cdot M^{-1}$), S represents the concentration of solute in the solid phase ($M \cdot M^{-1}$), D represents the hydrodynamic dispersion coefficient ($L^2 \cdot T^{-1}$), z represents the vertical spatial coordinate (L), and v represents the velocity of pore water ($L \cdot T^{-1}$), λ represents the rate of chemical reaction (T^{-1}), k_b represents the biodegradation rate (T^{-1}), and α represents first-order sorption coefficient (T^{-1}).

When the multi-process model considers the two kinetic adsorption (k_1, k_2-k_{d2}), biodegradation, and chemical reaction under microbe-mineral interaction conditions, the revised standard equations of convection, dispersion, adsorption, biodegradation, and chemical reaction can be written as

$$\frac{\partial C}{\partial t} + f\frac{\rho}{\theta}\frac{\partial S_1}{\partial t} + (1-f)\frac{\rho}{\theta}\frac{\partial S_2}{\partial t} = D\frac{\partial^2 C}{\partial z^2} - v\frac{\partial C}{\partial z} + \sum_k (k_{b_k} + \lambda_k)C, \quad (6-4)$$

$$\frac{\rho}{\theta}\frac{\partial S_1}{\partial t} = k_1 C, \quad (6-5)$$

$$\frac{\rho}{\theta}\frac{\partial S_2}{\partial t} = k_2 C - \frac{\rho}{\theta}k_{d2}S_2. \quad (6-6)$$

Where 1 and 2 represent the type-1 and type-2 sites, respectively, k_1 is the first-order sorption coefficient of site 1 (T^{-1}); k_2 is the first-order sorption coefficient of site 2 (T^{-1}), and k_{d2} is the desorption coefficient of site 2 (T^{-1}).

Sometimes, models that include two or more adsorption types are more realistic. Therefore, when the multi-process model considers linear adsorption (K_d), two kinetic adsorptions (k_1, k_2-k_{d2}), biodegradation, and chemical reaction under microbe-mineral interaction conditions, the revised standard equations of convection, dispersion, adsorption, biodegradation, and chemical reaction can be written as

$$\frac{\partial C}{\partial t} + f\frac{\rho}{\theta}\frac{\partial S_1}{\partial t} + (1-f)\frac{\rho}{\theta}\frac{\partial S_2}{\partial t} + (1-f)\frac{\rho}{\theta}\frac{\partial S_3}{\partial t}$$
$$= D\frac{\partial^2 C}{\partial z^2} - v\frac{\partial C}{\partial z} + \sum_k (k_{b_k} + \lambda_k)C, \quad (6-7)$$

$$\frac{\partial S_1}{\partial t} = K_d\frac{\partial C}{\partial t}, \quad (6-8)$$

$$\frac{\rho}{\theta}\frac{\partial S_2}{\partial t} = k_1 C, \quad (6-9)$$

$$\frac{\rho}{\theta}\frac{\partial S_3}{\partial t} = k_2 C - \frac{\rho}{\theta}k_{d2}S_2. \quad (6-10)$$

Where 1, 2, and 3 represent the type-1, type-2, and type-3 sites, respectively, K_d is the instantaneous equilibrium adsorption coefficient of site 1 ($L^3 \cdot M^{-1}$); k_1 is the first-order sorption coefficient of site 2 (T^{-1}); k_2 is the first-order sorption coefficient of site 3 (T^{-1}) and k_{d2} is the desorption coefficient of site 3 (T^{-1}).

(2) DMR model coupling with chemical reaction and biodegradation

The continuous-distribution multi-rate (CDMR) model combines the continuously distributed domain with the relevant sorption/desorption rate coefficients (Culver et al., 1997; Li and Brusseau, 2000; Saiers and Tao, 2000; Chen and Wagenet, 1995), which well represented the impact of media heterogeneity on sorption/desorption. The CDMR model can solve sorption/desorption hysteresis (Schnaar and Brusseau, 2014). The governing equations related to the model are defined as (Brusseau et al., 2019)

$$\frac{\partial C}{\partial t}+\frac{\rho_b}{\theta}\frac{\partial S_e}{\partial t}+\frac{\rho_b}{\theta}\sum_{i=1}^{m}f_i(k_{2i})\frac{\partial S_i}{\partial t}=-v\frac{\partial C}{\partial x}+D\frac{\partial^2 C}{\partial x^2}, \tag{6-11}$$

$$\frac{\partial S_e}{\partial t}=FK_fC_n\frac{\partial C}{\partial t}, \tag{6-12}$$

$$\frac{\partial S_i}{\partial t}=k_{2i}[(1-F)K_fC_n-S_i]. \tag{6-13}$$

where C represents the solvend concentration in the liquid phase ($M\cdot L^{-3}$), S_e represents the sorbed concentration in the solid phase in the instantaneous adsorption domain ($M\cdot M^{-1}$), S_i represents the sorbed concentration in the solid phase in the rate-limited adsorption domain ($M\cdot M^{-1}$), ρ_b represents the bulk density of the media ($M\cdot L^{-3}$), t represents the time, D represents the hydrodynamic dispersion coefficient ($L^2\cdot T^{-1}$), x represents the distance (L), v represents the average pore velocity ($L\cdot T^{-1}$), K_f represents the Freundlich adsorption isotherm coefficient ($L^3\cdot M^{-1}$), F represents the fraction of instantaneous adsorption domain, k_2 represents the first-order desorption rate constant (T^{-1}), m represents the total number of stagnant areas. The variable adsorption/desorption kinetics coefficient $k_2(T^{-1})$ was assumed to follow a non-normal distribution:

$$f(k_2)=\frac{1}{\sqrt{2\Pi}\,k_2\,\sigma_k}\exp\left(-\frac{[\ln(\ln k_2)-\mu]^2}{2\sigma_k^2}\right), \tag{6-14}$$

where μ is the average of $\ln k_2$; σ_k is the variance of $\ln k_2$.

When the multi-process model considers the chemical reaction and biodegradation under microbe-mineral interaction conditions, the revised standard equations of convection, dispersion, adsorption, chemical reaction, and biodegradation can be written as

$$\frac{\partial C}{\partial t}+\frac{\rho_b}{\theta}\frac{\partial S_e}{\partial t}+\frac{\rho_b}{\theta}\sum_{i=1}^{m}f_i(k_{2i})\frac{\partial S_i}{\partial t}=-v\frac{\partial C}{\partial x}+D\frac{\partial^2 C}{\partial x^2}+\sum_k(k_{b_k}+\lambda_k)C. \tag{6-15}$$

(3) TOSD model coupling with chemical reaction and biodegradation

The tempered one-sided stable density (TOSD) model can describe the comprehensive impacts of sorption, desorption, and diffusion mass transfer on the transport of contaminants in soil. Moreover, the TOSD-based physical model is also valid on a large scale. The governing equation of the model can be written as (Zhou et al., 2021)

$$\frac{\partial C_a}{\partial t}+\beta e^{-\lambda t}\frac{\partial^\alpha (e^{\lambda t}C_a)}{\partial t^\alpha}-\beta\lambda^\alpha C_a=-v\frac{\partial C_a}{\partial x}+D\frac{\partial^2 C_a}{\partial x^2}. \tag{6-16}$$

where C_a represents the solvent concentration in the liquid phase ($M \cdot L^{-3}$), t represents the time, D represents the hydrodynamic dispersion coefficient ($L^2 \cdot T^{-1}$), x represents the distance (L), and v represents the average pore velocity ($L \cdot T^{-1}$), α represents a time index, β represents the fractional capacity coefficient, and λ is the truncation parameter.

When the multi-process model considers the chemical reaction and biodegradation under microbe-mineral interaction conditions, the revised standard equations of convection, dispersion, adsorption, chemical reaction, and biodegradation can be written as

$$\frac{\partial C_a}{\partial t}+\beta e^{-\lambda t}\frac{\partial^\alpha (e^{\lambda t}C_a)}{\partial t^\alpha}-\beta\lambda^\alpha C_a=-v\frac{\partial C_a}{\partial x}+D\frac{\partial^2 C_a}{\partial x^2}+\sum_k(k_{b_k}+\lambda_k)C. \tag{6-17}$$

(4) Modified-TSM model coupling with chemical reaction and biodegradation

The partial differential equation governing the 1D nonequilibrium transport of contaminants during transient water flow in a variably saturated rigid porous medium is as follow (Silva et al., 2020):

$$\frac{\partial \theta C_w}{\partial t}+\rho_b\frac{\partial C_s}{\partial t}+\frac{\partial \Gamma A_{aw}}{\partial t}=\frac{\partial}{\partial x}\left(\theta D\frac{\partial C_w}{\partial x}\right)\frac{\partial v C_w}{\partial t}, \tag{6-18}$$

Where C_w, C_s, and Γ represent the solute concentration in the liquid ($M \cdot L^{-3}$), solid ($M \cdot M^{-1}$), and air-water interface ($M \cdot L^{-2}$) phases, respectively. θ represents the water content; ρ_b represents the bulk density of the porous medium ($M \cdot L^{-3}$); A_{aw} represents the area of air-water interface ($L^2 \cdot L^{-3}$); v represents the vector for the interstitial porewater velocity ($L \cdot T^{-1}$); D represents the hydrodynamic dispersion tensor ($L^2 \cdot T^{-1}$).

When the multi-process model considers the chemical reaction and biodegradation under microbe-mineral interaction conditions, the revised standard equations of convection, dispersion, adsorption, chemical reaction, and biodegradation can be written as

$$\frac{\partial \theta C_w}{\partial t} + \rho_b \frac{\partial C_s}{\partial t} + \frac{\partial \Gamma A_{aw}}{\partial t} = \frac{\partial}{\partial x}\left(\theta D \frac{\partial C_w}{\partial x}\right) - \frac{\partial v C_w}{\partial t} + \sum_{k}(k_{b_k} + \lambda_k) C. \tag{6-19}$$

(5) MPMT model coupling with biodegradation and chemical reaction

The multi-process rate-limited mass-transfer (MPMT) model is applied to depict the solvend transport affected by rate-limited solid-phase and air-water interface adsorption, preferential flow, and diffusive mass transfer in a system with non-advective regions. The standard equations of the model can be written as (Brusseau, 2020)

$$\frac{\partial C}{\partial t} + \rho_b \frac{\partial C_s}{\partial t} + \frac{\partial C_{aw}}{\partial t} = \frac{\partial}{\partial x}\left(D \frac{\partial C}{\partial x}\right) - \frac{\partial}{\partial x}(vC), \tag{6-20}$$

$$R_{a1} \frac{\partial C_a^*}{\partial t} + k_a^0 (C_a^* - S_a^*) + \omega (C_a^* - C_n^*) = \frac{1}{P} \frac{\partial^2 C_a^*}{\partial x^2} - \frac{\partial C_a^*}{\partial x}, \tag{6-21}$$

$$R_{n1} \frac{\partial C_n^*}{\partial t} + k_n^0 (C_n^* - S_n^*) = \omega (C_a^* - C_n^*), \tag{6-22}$$

$$R_{a2} \frac{\partial S_n^*}{\partial t} = k_a^0 (C_a^* - S_a^*), \tag{6-23}$$

$$R_{n2} \frac{\partial S_n^*}{\partial t} = k_n^0 (C_n^* - S_n^*). \tag{6-24}$$

where C is the aqueous concentration ($M \cdot L^{-3}$), C_s($M \cdot M^{-1}$) is the solid-phase adsorption, C_{aw} is the air-water interface adsorption $(M \cdot L^{-3})$, v is the vector for the interstitial porewater velocity $(L \cdot T^{-1})$, D is the hydrodynamic dispersion tensor $(L^2 \cdot T^{-1})$, ρ_b represents the bulk density of the porous medium $(M \cdot L^{-3})$, C_a^* is C_a/C_0, C_a is the solvend concentration in the advective region $(M \cdot L^{-3})$, C_0 is the input solute concentration $(M \cdot L^{-3})$, C_n^* is C_n/C_0, C_n is the solute concentration in the non-advective region $(M \cdot L^{-3})$, S_a^* is $S_{a2}/[(1-F_a)K_aC_0]$, S_{a2} is the proportion of the mass of adsorbate in the nonequilibrium adsorption phase to the mass of adsorbent in the advective region $(M \cdot M^{-1})$, F_a is the percentage of adsorbent in the instantaneous sorption in the advective region (–), K_a is the equilibrium adsorption constant for the advective region $(L^3 \cdot M^{-1})$, S_n^* is $S_{n2}/[(1-F_n)K_nC_0]$, S_{n2} is the proportion of the mass of adsorbate in the nonequilibrium adsorption phase to the mass of adsorbent in the non-advective region $(M \cdot M^{-1})$, F_n is the fraction of adsorbent in instantaneous sorption in the non-advective region (–), K_n is equilibrium adsorption coefficient in the non-advective region $(L^3 \cdot M^{-1})$, R_{a1} is the retardation of the equilibrium adsorption phase in the advective region, R_{a2} is the retardation of the nonequilibrium adsorption phase in the advective region, R_{n1} is the retardation of the equilibrium adsorption phase in the non-advective region, R_{n2} is the retardation of the nonequilibrium adsorption phase in the non-advective region, k_a^0 is the contribution of non-ideal adsorption in the advective region (–), k_n^0 is the contribution of non-ideal adsorption in the non-advective region (–), and ω is the contribution of diffusive mass transfer between the advective and non-advective domains (–).

Equations (21) and (22) are the mass balance of advective and non-advective regions respectively. Equations (23) and (24) are the nonequilibrium rate-limited adsorption for advective and non-advective regions, respectively.

When the multi-process model considers the chemical reaction and biodegradation under microbe-mineral interaction conditions, the revised standard equations of convection, dispersion, adsorption, chemical reaction, and biodegradation can be written as

$$\frac{\partial C}{\partial t} + \rho_b \frac{\partial C_s}{\partial t} + \frac{\partial C_{aw}}{\partial t} = \frac{\partial}{\partial x}\left(D \frac{\partial C}{\partial x}\right) - \frac{\partial}{\partial x}(vC) + \sum_k (k_{b_k} + \lambda_k) C. \quad (6-25)$$

(6) CCM model coupling with biodegradation and chemical reaction

The comprehensive compartment (CCM) model describes multiple processes for contaminant retention, namely, being sorbed in the solid phase, adsorption by non-aqueous phase liquid (NAPL), adsorption to the air-water interface, adsorption to the NAPL-water interface, adsorption to the air-NAPL interface, partitioning to NAPL, soil, and air. The governing equation of the model and the retardation factor R by all of the aforementioned processes are given as (Bahr and Rubin, 1987; Brusseau et al., 2019)

$$\frac{\partial C}{\partial t} + \rho_b \frac{\partial C_s}{\partial t} + \frac{\partial C'}{\partial t} + \frac{\partial}{\partial x}(vC) - \frac{\partial}{\partial x}\left(D \frac{\partial C}{\partial x}\right) = 0, \quad (6-26)$$

$$\frac{\partial C'}{\partial t} = \frac{\partial C_{aw}}{\partial t} + \frac{\partial C_{an}}{\partial t} + \frac{\partial C_{nw}}{\partial t}, \quad (6-27)$$

$$R = 1 + \frac{K_d \rho_b}{\theta_w} + \frac{K_a \theta_a}{\theta_w} + \frac{K_{aw} A_{aw}}{\theta_w} + \frac{K_n \theta_n}{\theta_w} + \frac{K_{nw} A_{nw}}{\theta_w} + \frac{K_{an} A_{an}}{\theta_w}. \quad (6-28)$$

Where C is the aqueous concentration ($M \cdot L^{-3}$), ρ_b is the bulk density of the porous medium ($M \cdot L^{-3}$), C_s is the solid-phase adsorption ($M \cdot M^{-1}$), C' is the adsorption at fluid-fluid interface ($M \cdot L^{-3}$), v is the vector for the interstitial porewater velocity ($L \cdot T^{-1}$), D is the hydrodynamic dispersion tensor ($L^2 \cdot T^{-1}$), C_{aw} is the air-water interfacial adsorption ($M \cdot L^{-3}$), C_{an} is the air-NAPL interfacial adsorption ($M \cdot L^{-3}$), C_{nw} is the NAPL-water interfacial adsorption ($M \cdot L^{-3}$), K_d represents the solid-phase sorption constant ($L^3 \cdot M^{-1}$), K_n represents the NAPL-water partition coefficient (–), K_a represents Henry's coefficient (–), K_{aw} represents the air-water interfacial adsorption constant ($L^3 \cdot L^{-2}$), K_{nw} represents the NAPL-water interfacial adsorption constant ($L^3 \cdot L^{-2}$), K_{an} represents the air-NAPL interfacial adsorption constant ($L^3 \cdot L^{-2}$), A_{aw} represents the specific air-water interfacial area ($L^2 \cdot L^{-3}$), A_{nw} represents the specific NAPL-water interfacial area ($L^2 \cdot L^{-3}$), A_{an} represents the specific air-NAPL interfacial area

($L^2 \cdot L^{-3}$), ρ_b represents the bulk density of the media ($M \cdot L^{-3}$), θ_a represents the volumetric air content (−), θ_n represents the volumetric NAPL content (−), and θ_w represents volumetric water content (−). $\theta_a + \theta_n + \theta_w = n$, where n represents porosity.

When the multi-process model considers the chemical reaction and biodegradation under microbe-mineral interaction conditions, the revised standard equations of convection, dispersion, adsorption, chemical reaction, and biodegradation can be written as

$$\frac{\partial C}{\partial t} + \rho_b \frac{\partial C_s}{\partial t} + \frac{\partial C'}{\partial t} + \frac{\partial}{\partial x}(vC) - \frac{\partial}{\partial x}\left(D \frac{\partial C}{\partial x}\right) - \sum_{k}(k_{b_k} + \lambda_k) C = 0. \tag{6-29}$$

6.2 Conclusions and future perspectives

The TSM is more likely used to describe chemical nonequilibrium actions, which is caused by the dynamic exchange adsorption and delayed adsorption of ions. The two-site adsorption can be combined by linear adsorption and single kinetic adsorption, or two kinetic adsorption. Sometimes, models can include two or more adsorption types, which consider linear adsorption, and two kinetic adsorptions to couple with biodegradation, and chemical reaction. The CDMR model combines the continuously distributed domain with the relevant sorption/desorption rate coefficients, which well represent the impact of media heterogeneity on sorption/desorption. The CDMR model can solve sorption/desorption hysteresis. The TOSD model can describe the comprehensive impacts of sorption, desorption, and diffusive mass transfer on the transport of contaminants in soil. Moreover, the TOSD-based physical model is valid on a large scale. The modified-TSM can be used to describe the transport of contaminants during transient water flow in a variably saturated rigid porous medium. The MPMT model can be applied to depict the solvend transport affected by rate-limited solid-phase and air-water interface adsorption, preferential flow, and diffusive mass transfer in a system with non-advective regions. The CCM model describes multiple processes for contaminant retention, namely, being sorbed in the solid phase, adsorption by non-aqueous phase liquid (NAPL), adsorption to the air-water interface, adsorption to the NAPL-water interface, adsorption to the air-NAPL interface, partitioning to NAPL, soil, and air.

Chapter 7
Some Other Physical-chemical Approaches Coupling Fe^0 Technology

7.1 WMF coupling with ZVI technology

7.1.1 Why does ZVI need a coupled WMF?

At the beginning of the 20th century, Faraday observed changes in the molecular environment, modifying characteristics and reactions, due to the interaction between external magnetic fields and intrinsic molecular magnetic properties, being called magnetochemistry. Later, a large number of experiments and theoretical insights on magnetic interactions in molecular scales have confirmed that the magnetic field can change the microstructure of the material, thus causing the change of properties of the material, and then affecting the chemical reaction. Therefore, magnetochemistry comes into being as a new discipline, which mainly focuses on studying the effect of magnetic fields on chemical reactions. Its main study objects include chemical kinetics, photochemistry, electrochemistry, synthesis reaction, polymerization reaction, isotope enrichment reaction, and biochemical reaction.

When ZVI is used for treating wastewater and contaminated groundwater, its system is often weak in selectivity to target pollutants (Noubactep, 2009), except for reacting with targeted pollutants, most of ZVI is consumed by oxidants, such as dissolved O_2, H_2O, and NO_3^- (Gu et al., 1999; Noubactep and Schoner, 2009). Therefore, large amounts of iron (hydr) oxides, such as lepidocrocite (γ-FeOOH), goethite (α-FeOOH), hematite (α-Fe_2O_3), maghemite (γ-Fe_2O_3), magnetite (Fe_3O_4), carbonate minerals (e.g. siderite ($FeCO_3$) and calcite

($CaCO_3$)), and others, could be produced in ZVI/H_2O systems, resulting in aging (passivation), which affects the efficiency and application of ZVI (Xu et al., 2016; Zhang et al., 2018). Numerous laboratory and field studies demonstrate that ZVI has a low intrinsic reactivity to contaminants due to its aging (passivation). The low reactivity of aged ZVI is an essential problem for the further development of ZVI-based technology (Xu et al., 2016), and it is vital to develop useful methods for significantly enhancing the reactivity of ZVI. Various techniques such as ultrasonic, acid washing, H_2-reducing pretreatment (Lai and Lo, 2008), electrochemical reduction (Chen et al., 2012), ZVI-based bimetals (Zhu and Lim, 2007), and nanosized ZVI (nZVI) (Huang et al., 2013) were used to improve ZVI reactivity. However, these methods always are complex, costly, and ecologically toxic (Guan et al., 2015).

Compared to the above methods, permanent magnets produced by WMF are energy-free and chemical-free (Xu et al., 2016b). The WMF is thus considered to be a promising and cutting-edge approach to improving ZVI reactivity. The advantages of WMF coupling with ZVI technology were shown in Figure 7.1.

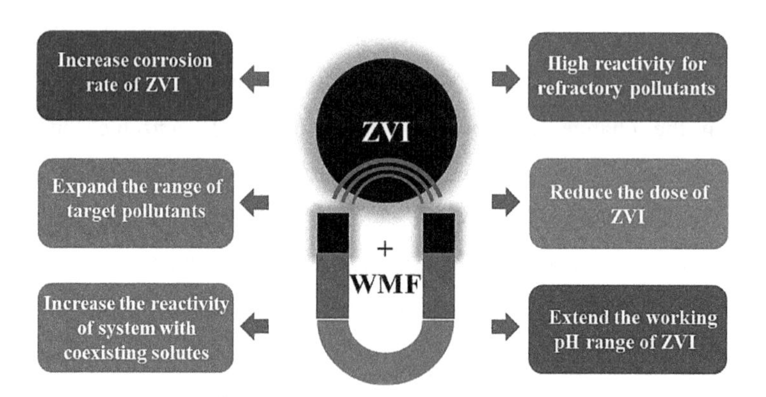

Figure 7.1 Advantages of WMF coupling with ZVI technology for remediation of groundwater and wastewater.

The WMF could significantly accelerate selenate (Se(VI)), As(V) (Li et al., 2015; Sun et al., 2017b) and selenite (Se(IV)) (Liang et al., 2014b) removals by ZVI. Apart from that, WMF-ZVI was applied to activate persulfate (PS) for treating wastewater contaminated with dyes and aromatic compounds

(Xiong et al., 2014). It was demonstrated that ZVI corrosion and Fe^{2+} release were promoted by WMF, which increased the sulfate radicals ($SO_4^- \cdot$) production and promoted the removal of diuron, benzotriazole, 4-nitrophenol, orange G, and caffeine (Xiong et al., 2014). WMF has also been used to accelerate the removal of Cu(II)/EDTA-Cu(II) (Jiang et al., 2015), Sb (V), Sb (III) (Fan et al., 2018; Xu et al., 2016b), and Cr(VI) (Li et al., 2017b) by ZVI. Because of the ferromagnetism of ZVI, the Lorentz force (F_L) and the magnetic gradient force ($F_{\Delta B}$) contributed positively to improving ZVI reactivity. Paramagnetic ions (e.g. Fe(II)) transport to the place of higher MF intensity under magnetic gradient force ($F_{\Delta B}$). As a result, corrosion products are locally distributed near the ZVI sphere and finally enhance ZVI reactivity (Zhang et al., 2018).

$$F_L = J \times B, \tag{7-1}$$

$$F_{\Delta B} = \left(\chi \times \frac{c}{\mu_0}\right) B(x) \frac{dB(x)}{dx}. \tag{7-2}$$

Hereby, J is the flux of charged species (coulombs $cm^{-2} \cdot s^{-1}$), B is an external magnetic field (T), χ is the molar magnetic susceptibility ($m^3 \cdot mol^{-1}$), $\mu_0 = 4\pi \times 10^{-7}$ $T \cdot mA^{-1}$, it is the magnetic permeability of vacuum, $B(x)$ is the MF intensity at position x (Li et al., 2017b; Xu et al., 2016b). The degradation of organic contaminants (chlorinated organic compounds, nitroaromatic compounds, dyes, etc) and sequestration of inorganic contaminants (heavy metals, arsenic, selenium, antimony, etc) by ZVI is controlled by the corrosion of ZVI, which is accompanied by the release of Fe^{2+}(Eq. 6) and the consumption of H^+. When the WMF or premagnetization is performed in the reaction systems, the change of ZVI interaction with organic and inorganic contaminants mainly comes from that the WMF or premagnetization could greatly enhance the corrosion of ZVI, the release of $\cdot OH$ (Eqs. 1 - 3), oxidation of Fe^{2+} and release of H^+ (Eqs. 7 - 9) (Li et al., 2015). The $\cdot OH$ could be produced more under WMF or premagnetization conditions. These $\cdot OH$ could facilitate the oxidation of organic and inorganic contaminants. In addition, some ZVI corrosion products (eg., γ-FeOOH) are favored for the adsorption of contaminants under WMF or premagnetization conditions (Li et al., 2015).

$$2Fe^0 + O_2 + 2H_2O \rightarrow 2Fe^{2+} + 4OH^-, \quad (7-3)$$

$$6Fe^{2+} + O_2 + 6H_2O \rightarrow 2Fe_3O_4 \downarrow + 12H^+, \quad (7-4)$$

$$4Fe^{2+} + O_2 + 6H_2O \rightarrow 4FeOOH \downarrow + 8H^+, \quad (7-5)$$

$$4Fe^{2+} + O_2 + 4H_2O \rightarrow 2Fe_2O_3 \downarrow + 8H^+. \quad (7-6)$$

Based on the previous studies, a mathematical model for describing the effect of WMF on the transport of paramagnetic ions in a solution was proposed (Waskaas and Kharkats, 1999; Waskaas and Kharkats, 2001). The model was established based on the magnetic gradient force ($F_{\Delta B}$) (Eq. 5), the solution flow, electrodiffusion, and solution electroneutrality are controlled by the Navier-Stokes equation (Eqs. 10 and 11), the Nernst-Planck equation (Eq. 12), and the Poisson equation (Eq. 13), respectively.

$$\frac{\partial \vec{\nu}}{\partial t} + (\vec{\nu}\nabla)\vec{\nu} = \vec{\nu}\nabla^2\vec{\nu} - \frac{\nabla P}{\rho} + \vec{g} + \frac{F_{\Delta B}}{\rho}, \quad (7-7)$$

$$(\nabla \vec{\nu}) = 0, \quad (7-8)$$

Where, $\vec{\nu}$ is the solution velocity, ρ is the solution density, g is the acceleration of gravity, and P is pressure.

Nernst-Planck equation for describing ionic transport of ith component in the solution:

$$(\nabla \vec{J}_i) = 0, \quad i = 1, 2, 3\cdots. \quad (7-9)$$

The Poisson equation provides the local electroneutrality of the solution in the model:

$$\sum_i z_i C_i = 0, \quad (7-10)$$

where, C and z are the ion concentration and number of corresponding electrical charges, respectively.

Based on the above model, $F_{\Delta B}$ could act on the paramagnetic ions (e.g. Fe^{2+}) and induce all components' convection in the solution (Waskaas and Kharkats, 1999; Waskaas and Kharkats, 2001).

7.1.2 Current status of WMF coupling with ZVI technology

Since the serendipitous invention of the WMF coupling with ZVI technology in 2014 (Guan et al., 2015), its ability to contaminants removal from groundwater and wastewater has been extensively investigated. The WMF was used to enhance ZVI to remove arsenic (As), selenium (Se), antimony (Sb), azo dye, 4-nitrophenol (4-NP), chromium (Cr), copper (Cu) and uranium (U), and all researches yielded a considerable effect. The promotion effect of WMF can be shown in the removal rate constant and removal efficiency. In Table 7.1, the contaminants that have been treated by WMF coupling with ZVI technology are presented.

The contaminants include organic contaminants, inorganic heavy metals, and metalloids. The kinetics of contaminants removal fit first-order: $\frac{dC}{dt} = -k_{obs}t$, the k_{obs} is the observed first-order rate constant. It showed that for Sb(V) with different initial concentrations, WMF increased the k_{obs} by 556% to 721% (Li, Jialiang. et al., 2015). The Cr(VI) removal also increased by 155% to 350% under WMF with ZVI condition (Feng et al., 2015). For Cu(II), when the initial pH is 3.0 - 6.0, the performance of removal increased by 989% to 37 400% under the WMF condition (Jiang et al., 2015). For orange G, the performance of removal increased by 539% to 2 804% under different pH conditions. The WMF also accelerated the decolorization rate and mineralization of orange G (Xiong et al., 2014). For 4-NP, the rate constant increased by 184% to 1 350% under different anion (Cl^-, SO_4^{2-}, NO_3^-, and ClO_4^-) conditions (Du et al., 2017). Since the beginning of the technology, two types of magnetization technology have been used; they are the pre-magnetization magnetic field and the permanent magnetic field. It was reported that the pre-magnetization method was adopted to improve the reactivity of passivated ZVI for removing high-concentration PNP (Ren et al., 2018), and the permanent magnetic field with ZVI was used to accelerate chloroacetamide removal in drinking water (Chen et al., 2019). Moreover, in the early stages of WMF coupling with ZVI technology, the reactive material ZVI only could treat a limited number of contaminants.

Table 7.1 The change of sequestration and degradation first-order rate constants (k_{obs}) of organic and inorganic contaminants in the WMF-ZVI system under different conditions.

Contaminant	Condition	Variable	k_{obs1}(min^{-1}) with WMF	k_{obs2}(min^{-1}) without WMF	k_{obs1}/k_{obs2}	Reference
Orange G	Experimental variable is pH, $C_{Orange\ G}$ = 0.2 mM, $C_{persulfate}$ = 2.0 mM, C_{ZVI} = 2.0 mM	pH=3	0.1732	0.0271	6.39	(Wang et al., 2000)
		pH=5	0.1452	0.0050	29.04	
		pH=7	0.0805	0.0052	15.48	
		pH=9	0.1129	0.0077	14.66	
Caffeine	pH_{ini} = 7, $C_{persulfate}$ = 0. 5 mM, C_{ZVI} = 0.5 mM	pH=7	0.0175	0.0021	8.33	
Benzotriazole			0.0126	0.0041	3.07	
Diuron			0.0291	0.0024	12.13	
4-nitrophenol	Experimental variable is concentration of anions, $C_{4\text{-nitrophenol}}$ = 10 μM, pH_{ini} = 4.0, C_{ZVI} = 0.1 g · L^{-1}	Anion (0 mM)	0.0130	0.0024	5.42	(Wang et al., 2000)
		Sulfate (1 mM)	0.0136	0.0047	2.89	
		Sulfate (10 mM)	0.0156	0.0055	2.84	
		Sulfate (50 mM)	0.0179	0.0061	2.93	
		Chloride (1 mM)	0.0123	0.0026	4.73	
		Chloride (10 mM)	0.0125	0.0041	3.05	
		Chloride (50 mM)	0.0142	0.0044	3.23	
		Nitrate (1 mM)	0.0127	0.0027	4.70	

(continued)

Contaminant	Condition	Variable	k_{obs1} (min^{-1}) with WMF	k_{obs2} (min^{-1}) without WMF	k_{obs1}/k_{obs2}	Reference
4-nitrophenol	Experimental variable is concentration of anions, $C_{4\text{-nitrophenol}}$ = 10 μM, pH_{ini} = 4.0, C_{ZVI} = 0.1 g · L^{-1}	Nitrate (10 mM)	0.0131	0.0026	5.04	(Wang et al., 2000)
		Nitrate (50 mM)	0.0145	0.0010	14.50	
		Perchlorate (1 mM)	0.0121	0.0019	6.37	
		Perchlorate (10 mM)	0.0114	0.0014	8.14	
		Perchlorate (50 mM)	0.0115	0.0014	8.21	
Cr(VI)	Experimental variable is C_{ZVI}, $C_{Cr(VI)}$ = 3.12 mg · L^{-1}, pH_{ini} = 5.0, ion strength 0.01 mol · L^{-1}	C_{ZVI} = 50 mg · L^{-1}	0.9000	0.2000	4.50	(Wang et al., 2017)
		C_{ZVI} = 100 mg · L^{-1}	2.8000	1.0000	2.80	
		C_{ZVI} = 200 mg · L^{-1}	2.8000	1.1000	2.55	
		C_{ZVI} = 300 mg · L^{-1}	2.6000	0.9000	2.89	
Cu(II)	Experimental variable is pH, $C_{Cu(II)}$ = 50 mg · L^{-1}, C_{ZVI} = 1.0 g · L^{-1}	pH = 3	0.2200	0.0202	10.89	(Wang and Wu, 2019b)
		pH = 6	0.1500	0.0004	375.00	
Sb(V)	Experimental variable is $C_{Sb(V)}$, reaction time 3 h, pH_{ini} = 4.0, C_{ZVI} = 1.0 g · L^{-1}	$C_{Sb(V)}$ = 5 mg · L^{-1}	0.1182	0.0144	8.21	(Phillips, 2009)
		$C_{Sb(V)}$ = 10 mg · L^{-1}	0.0929	0.0141	6.59	
		$C_{Sb(V)}$ = 20 mg · L^{-1}	0.0574	0.0086	6.67	
		$C_{Sb(V)}$ = 40 mg · L^{-1}	0.0177	0.0027	6.56	

(continued)

Contaminant	Condition	Variable	k_{obs1}(min^{-1}) with WMF	k_{obs2}(min^{-1}) without WMF	k_{obs1}/k_{obs2}	Reference
As(III)	Experimental variable is pH, $C_{As(III)}$ = 1 mg · L^{-1}, C_{ZVI} = 0.1 g · L^{-1}	pH = 3	6.4000	2.1000	3.05	(Bekele et al., 2019)
		pH = 5	2.9000	0.1200	24.17	
		pH = 7	2.1000	0.5300	3.96	
		pH = 9	2.5000	0.4000	6.25	
As(V)	Experimental variable is pH, $C_{As(V)}$ = 1 mg · L^{-1}, C_{ZVI} = 0.1 g · L^{-1}	pH = 3	7.5000	0.8000	9.38	(Li et al., 2021)
		pH = 5	5.8000	0.2000	29.00	
		pH = 7	5.0000	0.2800	17.86	
		pH = 9	5.0000	0.1500	33.33	
Se(IV)	Batch experiments, experimental variable is pH, $C_{Se(IV)}$ = 40 mg · L^{-1}, C_{ZVI} = 1.0 g · L^{-1}	pH = 4	2.1000	1.6000	1.31	(Park et al., 2021)
		pH = 5	2.7000	1.8600	1.45	
		pH = 6	1.5000	1.0000	1.50	
		pH = 7	1.6000	0.1000	16.00	
		pH = 7.2	1.6000	0.1000	16.00	

(continued)

Contaminant	Condition	Variable	k_{obs1}(min^{-1}) with WMF	k_{obs2}(min^{-1}) without WMF	k_{obs1}/k_{obs2}	Reference
Se(IV)	Experimental variable is $C_{Se(IV)}$, ion strength 0.01 mol · L^{-1}, pH_{ini} = 6.0, C_{ZVI} = 1.0 g · L^{-1}	$C_{Se(IV)}$ = 5 mg · L^{-1}	19.5000	1.2000	16.25	(Park et al., 2021)
		$C_{Se(IV)}$ = 10 mg · L^{-1}	15.5000	2.1500	7.21	
		$C_{Se(IV)}$ = 20 mg · L^{-1}	12.5000	1.7000	7.35	
		$C_{Se(IV)}$ = 40 mg · L^{-1}	1.3000	1.1000	1.18	
	Experimental variable is ageing time, ion strength 0.01 mol · L^{-1}, $C_{Se(IV)}$ = 40 mg · L^{-1}, C_{ZVI} = 1.0 g · L^{-1}	T = 0 h	14.5400	0.1080	134.63	(Lemic et al., 2021)
		T = 6 h	5.7500	0.3910	14.71	
		T = 12 h	4.9500	0.0928	53.34	
		T = 24 h	3.4000	0.0297	114.48	
		T = 48 h	3.8600	0.0352	109.66	
		T = 60 h	1.2900	0.0332	38.86	
		T = 80 h	0.013200	0.003990	3.31	
		T = 96 h	0.000306	0.000158	1.94	

Therefore, the WMF coupling with ZVI/H_2O_2 Fenton-like system (Xiang et al., 2016; Xiong et al., 2015), pre-magnetized ZVI/H_2O_2 Fenton-like system (Huang et al., 2018; Pan et al., 2016), WMF coupling with ZVI/persulfate system (Xiong et al., 2014) and pre-magnetized ZVI/persulfate system (Pan et al., 2017) were thus introduced to remove contaminants that were not amenable to the most frequently applied reactive material ZVI. With the introduction of these new Advanced Oxidation Processes (AOPs), it became possible to improve the reactivity of ZVI further to facilitate contaminant removal. Since then, a series of investigations have followed. Pre-magnetized ZVI/persulfate system and WMF coupling with ZVI/H_2O_2 Fenton-like system were used to treat groundwater contaminated with 4-nitrophenol (Xiong et al., 2015) and chlorinated solvents, such as 4-chlorophenol (Xiang et al., 2016) and 2,4-dichlorophenol (Li et al., 2017a). When these contaminants are effectively removed, their application was extended to sulfamethoxazole (Du et al., 2018), trichloroacetamide (Chen et al., 2019) and sulfamethazine (Pan et al., 2018a). The results of some of these investigations are prominent.

Therefore, the premagnetization of ZVI or WMF coupling with ZVI can be a suitable alternative technology to treat various pollutant-contaminated groundwater and wastewater. This book focuses on discussing and summarizing the reaction mechanisms and remediation performance of WMF coupling with ZVI and magnetic ZVI derivative technologies for different contaminants and applications.

7.2 WMF coupling with ZVI for treatment of contaminants

7.2.1 Heavy metals

Groundwater in uranium tailings ponds is frequently contaminated with Mn, As, U, Se, and V, some of which are presented in groundwater at production sites, base metal mining, weapons facilities and industrial sites (Morrison et al., 2002). The researchers used the low-cost reactive materials containing powdered ZVI or granular ZVI to build treatment cells of permeable reactive barriers (PRBs) and found that the concentrations of metal ions As, Mn, etc. significantly reduced in groundwater that flowed through PRBs (Morrison et al., 2002). However, the passivation of ZVI seriously influences the long-term running of the system. Many researchers have adopted magnetic materials or WMF coupling with ZVI technology for remediation of Cd(II), Cr(IV), Cr(VI), Pb(II), and Cu(II) contaminated water (Feng et al., 2015; Huang et al., 2013; Jiang et al., 2015; Li et al., 2017b; LYU et al., 2014).

The non-electrosprayed nanometer ZVI (NE-nZVI) and electrosprayed nanometer ZVI (E-nZVI) were applied to treating wastewater contaminated with Pb(II), Cr(IV) and Cd(II) under magnetic separation conditions (Huang et al., 2013). It was found that over 80% of Pb(II), Cr(IV) and Cd(II) were removed by NE-nZVI or E-nZVI. In addition, with increasing the treatment time of magnetic, removal of Pb(II) improved by 20%. Furthermore, magnetic treatment significantly reduced the nZVI dosage. Therefore, Pb(II), Cr(IV), and Cd(II) were quickly removed from wastewater via nZVI with magnetic treatment (Huang et al., 2013). Moreover, the nZVI was assembled on magnetic Fe_3O_4/graphene to sequestrate Cr(VI). It was demonstrated that the magnetic Fe_3O_4/graphene had excellent removal efficiency (Table 7.2) and capacity for Cr(VI) and could persistently maintain the nZVI corrosion. Therefore, the nZVI could continuously release electrons on its surface to reduce Cr(VI) to Cr(III) (Lv et al., 2014). Furthermore, Feng et al (2015) found that WMF coupling with ZVI could accelerate the sequestration of Cr(VI) but did not change the removal mechanism. It came that the change of the apparent activation energy of Cr(VI) sequestration in WMF-ZVI and ZVI systems was small (Feng et al., 2015).

Li et al. (2017b) explored the effect of magnetic field intensity on the removal of Cr(VI) by different particle sizes of ZVI to determine the relative contribution of F_L and $F_{\Delta B}$ for WMF enhancing effect (Li et al., 2017b). To a certain extent, the greater the magnetic field strength, the better the removal effect of Cr(VI). However, the WMF up to 10.0 mT shows a negligible effect on Cr(VI) sequestration by non-ferromagnetic Zn(0), suggesting that the F_L has a minor role in it. Therefore, $F_{\Delta B}$ was considered to be the main driving force for removing Cr(VI) under ZVI coupling with the WMF condition (Figure 7.2) (Li et al., 2017b). The WMF coupling with ZVI also was applied for removing

Table 7.2 Comparison of the performance of different ZVI derivative technologies.

	ZVI derivative technology	Contaminant	Reaction Condition	Time (min)	k_{obs} (min^{-1})	Removal efficiency (%)	Reference
Heavy metals	nZVI-magnetic Fe_3O_4/graphene	Cr(VI)	$C_{Cr(VI)}$ = 10.0 mg · L^{-1}, graphene (0.09 g), nZVI (0.02 g), Fe_3O_4 NPs (0. 30 g), nZVI @ MG (0.41 g), pH = 10	120	—	83.8	Lv et al., 2014
	nZVI				—	18.0	
	magnetic Fe_3O_4				—	21.6	
	graphene				—	23.7	
Chlorinated organic compounds	AC electromagnetic field-nZVI	TCE	C_{TCE} = 50. 0 mg · L^{-1}, C_{ZVI} = 10.0 g · L^{-1}, AC EMF at the current intensity of 15 A and the frequency of 150 kHz.	60	0.0352	87.9	Phenrat et al., 2016
	nZVI				0.0071	34.7	
	Pre-magnetized ZVI/PS	2,4-DCP	$C_{2,4\text{-}DCP}$ = 4. 0 mg · L^{-1}, C_{ZVI} = 1.0 mM, $C_{persulfate}$ = 1. 0 mM and pH = 10	60	0.1070	98.6	Li et al., 2017a
	ZVI/PS				0.0125	52.7	
	Pre-magnetized ZVI/PS	4-CP	$C_{4\text{-}CP}$ = 20 mg · L^{-1}, C_{ZVI} = 1.0 mM, $C_{persulfate}$ = 1. 0 mM and pH = 7	60	0.1120	99.9	Li et al., 2017b
	ZVI/PS				0.0115	49.8	
	Pre-magnetized ZVI/PS	CB	C_{CB} = 20 mg · L^{-1}, C_{ZVI} = 1.0 mM, $C_{persulfate}$ = 1.0 mM and pH = 7	60	0.0330	86.0	Li et al., 2017b
	ZVI/PS				0.0085	40.0	

(**continued**)

	ZVI derivative technology	Contaminant	Reaction Condition	Time (min)	k_{obs} (min^{-1})	Removal efficiency (%)	Reference
Chlorinated organic compounds	Pre-magnetized ZVI-C/PS	2,4-D	$C_{2,4\text{-}D}$ = 20 mg · L^{-1}, $C_{ZVI\text{-}C}$ = 0.05 g · L^{-1}, $C_{persulfate}$ = 1.0 mM and pH = 3	30	0.0680	87.0	Li et al., 2018
	ZVI-C/PS				0.0310	60.5	
	WMF-ZVI/H_2O_2	4-CP	$C_{4\text{-}CP}$ = 50 mg · L^{-1}, C_{ZVI} = 0.1 g · L^{-1}, C_{H2O2} = 1.0 mM and pH = 3	20	0.1600	95.9	Xiang et al., 2016
	ZVI/H_2O_2				0.0180	30.6	
	Pre-magnetized ZVI/H_2O_2	2,4-DCP	$C_{2,4\text{-}DCP}$ = 4.0 mg · L^{-1}, C_{ZVI} = 2.0 mM, C_{H2O2} = 1.0 mM and pH = 4	30	0.1290	97.9	Pan et al., 2016
	ZVI/H_2O_2				0.0370	67.0	
	WMF/ZVI/EDTA	Diclofenac	$C_{Diclofenac}$ = 10.0 mg · L^{-1}, C_{ZVI} = 0.4 g · L^{-1}, C_{EDTA} = 2.0 mM and pH = 5	120	0.0250	95.0	Zhou et al., 2018
	ZVI/EDTA				0.0070	56.8	
Nitroaromatic compounds	WMF-ZVI/PS	PNP	$C_{PNP=}$ 0.02 mM, C_{ZVI} = 0.5 mM, $C_{persulfate}$ = 0.5 mM and pH = 7	60	0.0700	98.5	Xiong et al., 2014
	ZVI/PS				0.0063	31.5	
	WMF-ZVI/H_2O_2	PNP	C_{PNP} = 0.02 mM, C_{ZVI} = 0.5 mM, C_{H2O2} = 0.5 mM and pH = 4	60	0.0700	99.0	Xiong et al., 2015
	ZVI/H_2O_2				0.0200	60.0	
	Pre-magnetized ZVI/PS	PNP	C_{PNP} = 20 mg · L^{-1}, C_{ZVI} = 1.0 mM, $C_{persulfate}$ = 1.0 mM and pH = 7	60	0.0420	92.0	Li et al., 2017b
	ZVI/PS				0.0038	20.0	
	Pre-magnetized ZVI/PS	*p*-nitrochlorobenzene	$C_{p\text{-}nitrochlorobenzene}$ = 20 mg · L^{-1}, C_{ZVI} = 1.0 mM, $C_{persulfate}$ = 1.0 mM and pH = 7	60	0.0490	95.0	
	ZVI/PS				0.0153	60.0	

(continued)

	ZVI derivative technology	Contaminant	Reaction Condition	Time (min)	k_{obs} (min^{-1})	Removal efficiency (%)	Reference
Dyes	WMF-ZVI/PS	Orange G	C_{orange} G = 0.2 mM, C_{ZVI} = 2.0 mM, $C_{persulfate}$ = 2.0 mM and pH = 3	10	0.3800	98.0	Xiong et al., 2014
	ZVI/PS				0.0590	44.6	
	Pre-magnetized ZVI/PS	RhB	C_{RhB} = 100 $mg \cdot L^{-1}$, C_{ZVI} = 4.0 mM, $C_{persulfate}$ = 8.0 mM and pH = 7	10	0.2330	90.3	Pan et al., 2018c
	ZVI/PS				0.0940	60.9	
Others	Pre-magnetized ZVI/PS	Phenol	C_{Phenol} = 20 $mg \cdot L^{-1}$, C_{ZVI} = 1.0 mM, $C_{persulfate}$ = 1.0 mM and pH = 7	60	0.0480	94.0	Li et al., 2017b
	ZVI/PS				0.0140	57.0	
	DC magnetically cast-iron scraps fixed bed	SO_2	C_{SO2} = 1 200 ppm, Q_L/Q_G(the ratio of volumetric liquid flow rate and gas flow rate) was kept 1 $L \cdot m^{-3}$, Q_G = 10 $L \cdot min^{-1}$	180	—	80.0	Jiang et al., 2008
	Cast-iron scraps fixed bed				—	50.0	
	Pre-magnetized ZVI/H_2O_2	Salty wastewater	Reverse osmosis concentrated wastewater, C_{ZVI} = 16. 0 mM, C_{H2O2} = 8.0 mM and pH = 3	60	0.0380	83.8	Pan et al., 2018b
	ZVI/H_2O_2				0.0124	52.4	
	Pre-magnetized ZVI/H_2O_2		Petrochemical wastewater, C_{ZVI} = 320.0 mM, C_{H2O2} = 16.0 mM and pH = 3		0.0290	82.6	
	ZVI/H_2O_2				0.0092	42.5	

(continued)

	ZVI derivative technology	Contaminant	Reaction Condition	Time (min)	k_{obs} (min^{-1})	Removal efficiency (%)	Reference
Others	Pre-magnetized ZVI/H_2O_2	Citric acids wastewater	Real wastewater, C_{COD} = 145 mg · L^{-1}, initial pH = 3. 0, H_2O_2/COD= 0. 75, H_2O_2/COD = 3. 29, precipitation pH = 9.0	120	0.0057	49.2	Huang et al., 2018
	ZVI/H_2O_2				0.0037	35.5	
	US/pre-magnetized ZVI/persulfate	SMT	C_{SMT} = 5 mg · L^{-1}, C_{ZVI} = 0.1 mM, $C_{persulfate}$ = 1.0 mM, US = 60 W and pH = 7	60	0.0459	94.0	Pan et al., 2018a
	US/ZVI/persulfate				0.0122	52.0	
	pre-magnetized ZVI/persulfate				0.0119	51.0	
	UV/pre-magnetized ZVI/H_2O_2	SMT	Real wastewater, C_{ZVI} = 0. 3 mM, C_{H2O2} = 0. 3 mM, UV = 6 W and pH = 7.3	30	0.1694	92.1	Pan et al., 2019a
	UV/ZVI/H_2O_2				0.0941	72.1	
	UV/H_2O_2				0.0689	53.9	
	UV/pre-magnetized ZVI/persulfate	SMT	Real wastewater, C_{SMT} = 0.4 mg · L^{-1}, C_{ZVI} = 0. 1 mM, $C_{persulfate}$ = 0. 2 mM, UV = 6 W and pH = 7.3	20	0.3008	99.8	Pan et al., 2019b
	UV/ZVI/PS				0.1431	94.3	
	UV/PS				0.0923	84.2	

paramagnetic Cu(II), which enhanced the Cu(II) adsorption and promoted electron transfer. Due to the uneven distribution of magnetic field strength, the $F_{\Delta B}$ accelerated the Cu(II) transport to the ZVI surface. It was shown that WMF can accelerate the ZVI corrosion and facilitate Cu(II) reduction for that magnetic increased mass transfer (Jiang et al., 2015).

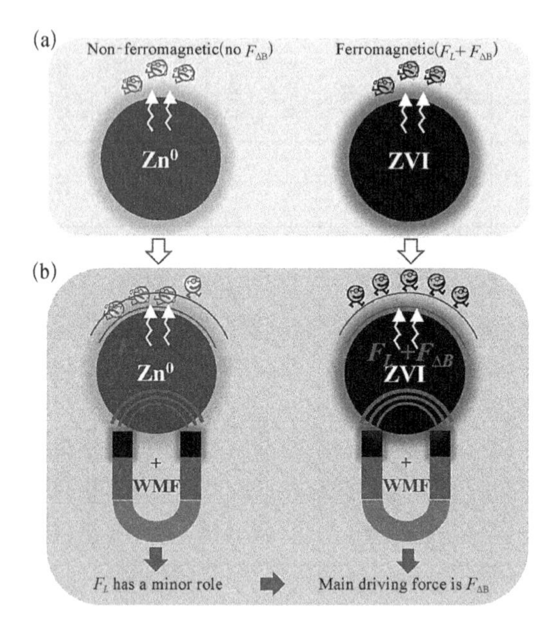

Figure 7. 2 Relative contribution of Lorentz force (F_L) and magnetic gradient force ($F_{\Delta B}$) for WMF enhancing effect on the sequestration of Cr(VI) (Wang et al., 2000).

In summary, magnetic materials or WMF coupling with ZVI was a novel technology for treating heavy metals contaminated wastewater. It could accelerate the ZVI corrosion, promote the electron transfer, and accelerate the heavy metal ions transport to the ZVI surface.

7.2.2 Chlorinated organic compounds

Chlorinated organic compounds (COCs), such as 2, 4, 6-trichlorophenol (2,4,6-TCP), 2, 4-dichlorophenol (2, 4-DCP), 4-chlorophenol (4-CP), trichloroethylene (TCE), tetrachloroethylene (PCE), 1, 1, 1-trichloroethane (1,1,1-TCA), polychlorinated biphenyls (PCB) and 1, 1, 1-trichloro-2, 2-bis

(4-chlorophenyl) ethane (DDT) are implemented on dyeing, pharmaceutical, electronic, chemical and other industries (Song et al., 2016; Wang and Wu, 2017; Xiang et al., 2016; Yang et al., 2018). They are the most common COCs in wastewater and groundwater and can pose threats to animals, plants and humans due to carcinogenesis and bioaccumulation (He et al., 2010; Huang et al., 2014). ZVI-based techniques are widely applied for the removal of COCs because ZVI is non-toxic, cheap, and reactive (Mdlovu et al., 2019; Wang and Wu, 2017). However, ZVI techniques have some limitations in treating refractory organic contaminants (Guan et al., 2015), including reactivity loss with time from the metal hydroxides and precipitation, low reactivity from the passive layer of ZVI, narrow working pH range, low removal efficiency for some refractory COCs and so on. Recently, WMF is used to resolve the above problems. The WMF-ZVI system (Chen et al., 2019; Zhou et al., 2018), WMF-ZVI/PS system (Xiong et al., 2014), pre-magnetized ZVI/PS system (Li et al., 2017a), WMF-ZVI/H_2O_2 system (Kim et al., 2011), pre-magnetized ZVI/H_2O_2 system (Pan et al., 2016) and pre-magnetized ZVI-C/PS system (Li et al., 2018) were developed for remediation of COCs contaminated water. A large number of COCs, including TCE, triclosan, trichloroacetamide (TCAM), 2, 4-DCP, 2, 4-dichlorophenoxyacetic acid (2,4-D), 4-CP, chlorobenzene (CB), and diclofenac have been successfully removed in these systems.

It was reported that the dechlorination of TCE increased by 496% using ZVI and nZVI with a 150 kHz alternating current (AC) electromagnetic field (Table 7.2). The results demonstrated that the electromagnetic field also could accelerate ZVI corrosion and facilitate TCE removal (Phenrat et al., 2016). In addition, it was shown that the triclosan removal was significantly increased in the WMF-ZVI system under acidic conditions. Compared to anaerobic conditions, the system had a greater removal rate under aerobic conditions (Wu et al., 2018). The WMF-ZVI system also presented a much better performance for TCAM removal than the ZVI system and the enhanced effect was more significant under higher pH or lower ZVI dose conditions (Chen et al., 2019). Moreover, the WMF enhanced the O_2 transport to the ZVI surface and promoted TCAM hydrolysis. The magnetic field was also used to induce ZVI oxidative reactions for 4-CP removal (Kim et al.,

2011). They found that the WMF significantly increased the 4-CP degradation from 26% to 54% due to the improved generation of · OH resulting from the more accessible O_2 transport to the ZVI surface under magnetic field conditions.

Also, a novel process of activating persulfate with ZVI under WMF conditions (WMF-ZVI/PS system) was utilized to degrade 2,4-DCP and 4-CP (Xiong et al., 2014). This research revealed that WMF increased the production of sulfate radicals (SO_4^{-} ·) and facilitated the removal of 2,4-DCP and 4-CP. It was from that WMF significantly improved the release of Fe^{2+}. Li et al (2017) adopted the pre-magnetization process to remove 2,4-DCP, 4-CP, and CB by activated persulfate with ZVI. The results demonstrated that the performance of the pre-magnetized ZVI/PS system was 1.87 – 5.18 times that of the ZVI/PS system (Table 7.2), which confirmed the effectiveness of pre-magnetization in promoting ZVI corrosion and thus promoting 2,4-DCP, 4 – CP and CB removal (Li et al., 2017a; Li et al., 2017b). The inexpensive ZVI-C after pre-magnetization also was used to activate persulfate (pre-magnetized ZVI-C/PS system) for 2, 4-D degradation (Li et al., 2018). The pre-magnetized ZVI-C/PS system significantly improved the degradation and de-chlorination of 2,4-D. Compared with the pre-magnetized ZVI/PS system (Table 7.2), the pre-magnetized ZVI-C/PS system had much better stability and reusability. Furthermore, The removal performance of WMF-ZVI/H_2O_2 for 4-CP was evaluated (Xiang et al., 2016). The WMF accelerated the Fenton reactions to produce · OH faster. Apart from that, WMF modified the surface of pristine ZVI and improved the mass transfer of the reactions.

The degradation performance of 2,4-DCP in pre-magnetized ZVI/H_2O_2 and ZVI/H_2O_2 systems also were investigated (Pan et al., 2016). The degradation rate of 2,4-DCP in the pre-magnetized ZVI/H_2O_2system was 3.49 – 5.6 times that in the ZVI/H_2O_2 system (Table 7.2). The "magnetic memory" of pre-magnetized ZVI can be used for 24 hours, which is suitable for off-site use. After the second pre-magnetization, part or all of the "magnetic memory" can be recovered. It was reported that a magnetic field/ZVI/EDTA Fenton-like system was developed to degrade diclofenac (Zhou et al., 2018). The results suggested that the magnetic field does not alter the Fenton-like reactions and the homogeneous iron cycle. However, it accelerated the ZVI surface corrosion and altered the heterogeneous ZVI surface-bond reactions (Figure 7.3).

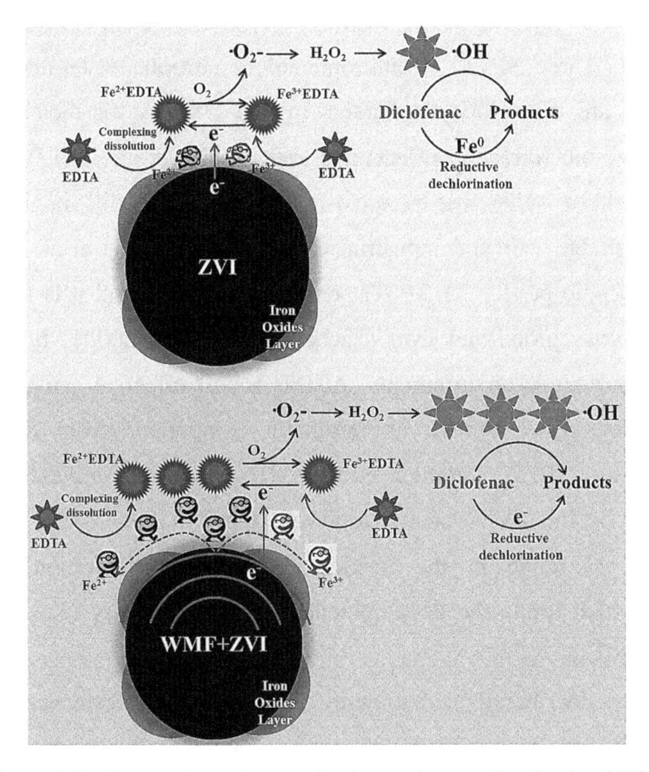

Figure 7.3 Proposed reaction mechanism and promotional role of WMF in the WMF/ZVI/EDTA Fenton-like system (Wang et al., 2000).

In general, WMF coupling with ZVI for removing COCs from wastewater and groundwater is a potentially effective technology, since it is energy-free and chemical-free. Despite the increased application and development of WMF-ZVI and magnetic ZVI derivative technologies since 2014 (Guan et al., 2015), research about COCs removal by WMF-ZVI and magnetic ZVI derivative technologies is still limited, and the exact mechanisms have not been conclusively established. Of course, further optimization and studies are necessary.

7.2.3 Nitroaromatic compounds

The increasing pollution of nitroaromatic compounds in water has caused widespread concern for their adverse effects on water ecology and human health (Du et al., 2017). Especially nitrophenols (NPs), a relevant raw material, have been extensively used in drug, herbicide, fungicide, paint, rubber and other chemical industries (Arora, 2012; Arora et al., 2012; Arora et al., 2014; Ju and Parales, 2010). Three NPs (2,4-dinitrophenol, 4-nitrophenol (*p*-nitrophenol) and 2-nitrophenol) are the priority pollutants by the USEPA for their mutagenicity, carcinogenicity, bio-refractory effects and toxicity (Du et al., 2017). As a priority NP, 4-nitrophenol (4-NP) was detected in groundwater, soil, rainwater, surface water, active sludge, air and industrial effluents (Inglezakis et al., 2017; Preiss et al., 2009; Rubio et al., 2012; Watson et al., 2017), and it is very hazardous to central nervous, blood and liver (Eichenbaum et al., 2009). It is essential to control 4-NP in aquatic environments. ZVI as a cost-effective active material was extensively adopted to remove nitroaromatic compounds from the wastewater (Mukherjee et al., 2015; Stefaniuk et al., 2016). The derivatives of 4-NP could also be treated by the ZVI system. However, the ZVI system has low efficiency and a harsh pH range in the treatment of high-concentration nitroaromatic compounds, which limits the development of this technology (Lai et al., 2014; Zhu and Ni, 2011).

Therefore, ZVI-based technologies with better performance need to be developed for the remediation of nitroaromatic compounds contaminated wastewater. Here, WMF combined with the ZVI system was used to enhance the removal of nitroaromatic compounds. The WMF-ZVI system, WMF-ZVI/PS system, pre-magnetized ZVI/PS system, and WMF-ZVI/H_2O_2 system were developed to treat 4-nitrophenol (*p*-nitrophenol or PNP) (Du et al., 2017; Ren et al., 2018; Xiong et al., 2014; Xiong et al., 2015) and *p*-nitrochlorobenzene (Li et al., 2017b) contaminated water. The degradation of PNP by WMF with six kinds of ZVI was studied (Du et al., 2017). It was shown that the first-order rate constants k_{obs} of PNP degradation in the WMF-ZVI system were 2.9 – 5.4 times greater than that in alone ZVI system. The enhancement of PNP removal by WMF-ZVI mainly came from the improvement of ZVI corrosion. The WMF significantly accelerated the PNP removal, especially at pH = 7. In addition, the k_{obs} of PNP improved with the increase of magnetization time.

The WMF-ZVI/PS system was also utilized to degrade PNP. The results showed that the degradation rate of PNP by ZVI/PS improved 11.1 times under the WMF condition (Table 7.2) (Xiong et al., 2014). It revealed that WMF facilitated the Fe^{2+} release, which induced and improved $SO_4^- \cdot$ production and thus promoted the PNP degradation (Figure 7.4). In addition, Xiong et al (2015) studied the enhancement of the WMF coupled with the advanced Fenton process (WMF-ZVI/H_2O_2) for PNP degradation (Xiong et al., 2015). It was found that the PNP degradation significantly improved in the WMF-ZVI/H_2O_2 system at an initial pH of 3.0 to 6.0. The cumulative amount of $\cdot$OH in the WMF-ZVI/H_2O_2 system was 3 times that in the ZVI/H_2O_2 system and the $\cdot$OH was the primary oxidant for the removal of PNP. The WMF improved $\cdot$OH production by facilitating the ZVI corrosion and Fe^{2+} release (Figure 7.4), which was the key and limiting step in the ZVI/H_2O_2 system.

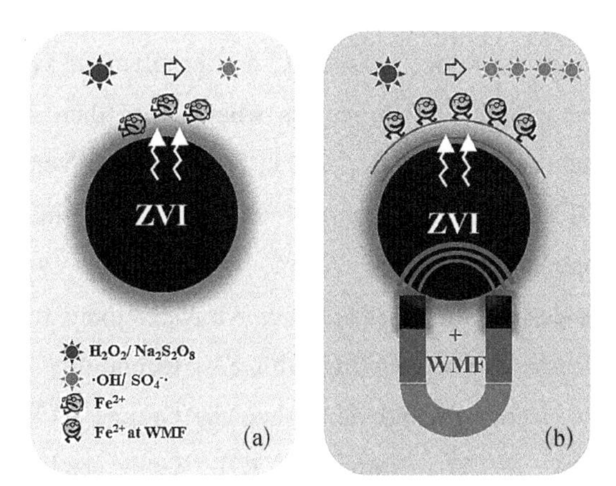

Figure 7.4 Change mechanism of hydroxyl radicals ($\cdot$OH) or sulfates radicals ($SO_4^- \cdot$) in a WMF-ZVI/H_2O_2 system or a WMF-ZVI/PS system. (a) and (b) represent the systems without or with WMF, respectively (Amoako-Nimako et al., 2021; Xiong et al., 2014).

The pre-magnetization method was adopted to improve and maintain the reactivity of ZVI/(passivated ZVI) for treating high-concentration (500 mg $\cdot$ L^{-1}) PNP wastewater (Ren et al., 2018). It revealed that the pre-magnetized ZVI/(passivated ZVI) system obtained much greater k_{obs} for PNP removal. The passivated ZVI was stimulated better than ZVI in the pre-magnetization process. Li et al. found that the removal rates of *p*-nitrochlorobenzene and PNP were significantly improved in the pre-magnetized ZVI/PS system, and the k_{obs} were 10.4 and 2.76 times that in ZVI/PS system, respectively (Li et al., 2017). Pre-magnetized ZVI combined with AOPs technology accelerated the degradation of PNP and *p*-nitrochlorobenzene. Compared with other alone AOPs (electrochemical oxidation, photo-Fenton, etc.), pre-magnetized ZVI has a more potent effect on improving the reaction rate of ZVI/PS process (Li et al., 2017b).

Based on the views above, WMF coupling with ZVI and magnetic ZVI combined with AOPs derivative technologies for removing nitroaromatic compounds from wastewater and groundwater are potentially practical and environmentally friendly technologies. They could reduce the strong resistance of NPs to biological and chemical oxidation.

7.2.4 Dyes

Reactive azo dyes are toxic, persistent, and problematic synthetic compounds that are widely used in industrial processes where more than 20% of dyes reach industrial wastewater (Bafana et al., 2011; de Campos Ventura-Camargo and Marin-Morales, 2013). Various treatment technologies of industrial wastewater have been developed (Singh and Arora, 2011). However, many of the technologies do not destroy the dyes but rather transfer them from one phase into another. Nowadays, the WMF coupling with ZVI technology is proven to be an attractive solution to decolorize and mineralize the Orange I (Xu et al., 2016a), Orange II (Xiao et al., 2014), Orange G (OG) (Xiong et al., 2014), amaranth (AR27) (Feng et al., 2015) and Rhodamine B (RhB) (Pan et al., 2018c) from industrial wastewater.

The Orange I removal by micron-sized granular ZVI with magnetization by WMF and sulfidation by pretreatment with sulfide was studied (Xu et al., 2016a). The results showed that both magnetization and sulfidation increased the removal rates of Orange I by 2.4 to 71.8 times. Moreover, it was shown that the k_{obs} of Orange II removal in the WMF-ZVI system increased by 260% to 2 100%. The improvement of Orange II removal came from the WMF accelerated the ZVI corrosion (Xiao et al., 2014). The pre-magnetization approach was also adopted to increase the removal rate of dyes. Liang et al. adopted the pre-magnetization to boost the reactivity of multiple ZVI toward AR27 (Feng et al., 2015). Compared with the ZVI without pre-magnetization, all pre-magnetized ZVI had higher intrinsic reactivity for AR27 removal.

WMF and ZVI also were adopted to activate PS (WMF-ZVI/PS) synergistically to remove OG (Xiong et al., 2014). It was shown that the WMF-ZVI/PS system induced a 540% to 2 820% enhancement for OG removal under different initial pH, ZVI and PS dosage conditions. The k_{obs} of OG in the WMF-ZVI/PS system increased with decreasing initial pH or increasing ZVI or PS dosages. The WMF significantly improved the mineralization and the decolorization rate of OG. The $SO_4^- \cdot$ was the primary free radical and played the primary role in OG removal in the WMF-ZVI/PS system. Pan et al (2018c) studied the mechanism and interferences of anions (CO_3^{2-}, HCO_3^-, PO_4^{3-}, NO_3^-, HPO_4^{2-}, SO_4^{2-}, Cl^-) and cations (Ca^{2+}, Mn^{2+}, Cu^{2+}, Mg^{2+}) for RhB removal in pre-magnetized ZVI/PS system (Figure 7.5) (Pan et al., 2018c). It

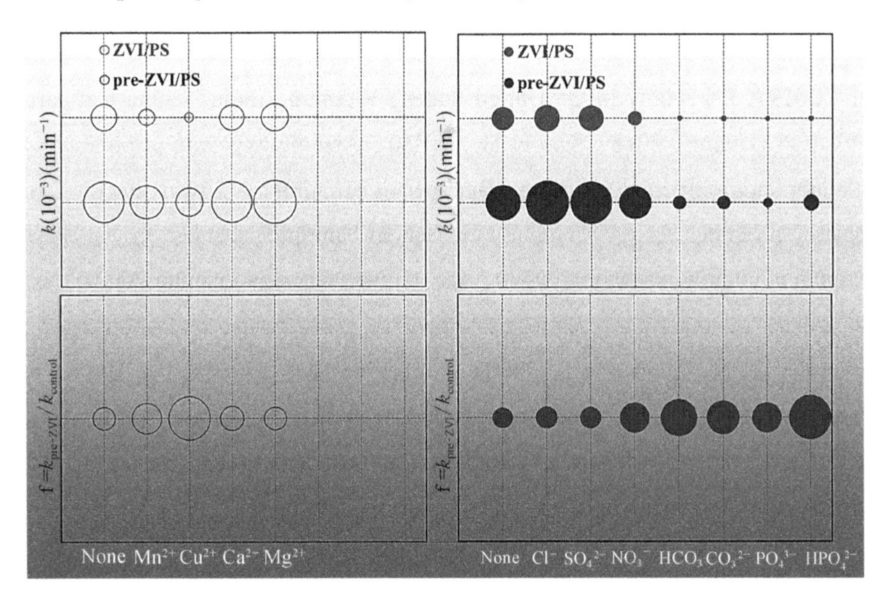

Figure 7.5 First-order rate constants (k_{obs}) and f ($= k_{pre\text{-}ZVI}/k_{conrol}$) in ZVI/PS system (Control) and pre-magnetized ZVI/PS system under different anions (CO_3^{2-}, HCO_3^-, PO_4^{3-}, NO_3^-, HPO_4^{2-}, SO_4^{2-} and Cl^-) and cations (Ca^{2+}, Mn^{2+}, Cu^{2+} and Mg^{2+}) conditions (Xu et al., 2016).

revealed that the removal rate of RhB in the pre-magnetized ZVI/PS system was 2.1 – 37.5 folds that in the ZVI/PS system (Control). The role of $SO_4^- \cdot$ generated in the reaction of PS with ZVI in the system. It was found that the pre-magnetized ZVI/PS system exhibited many advantages over other systems without WMF for the removal of RhB in terms of chemical dose, working pH range, and removal efficiency.

Taken together the arguments, the application of WMF-ZVI, WMF-ZVI/PS, and pre-magnetized ZVI/PS are cost-effective approaches for degrading Orange I, Orange II, OG, AR27, and RhB. They can improve the removal efficiency, extend the working pH range and reduce the chemical dose of ZVI and PS.

7.2.5 Arsenic

Arsenic (As) is an ubiquitous and highly toxic metalloid (Qiu et al., 2017; Ren et al., 2017; Ren et al., 2019) and As contamination occurs widely in surface and groundwater, which affects water quality and human health (Fendorf et al., 2010; Rodriguez-Lado et al., 2013). Because of its high toxicity, limits on As concentrations in potable water are meager (10 $\mu g \cdot L^{-1}$ to 50 $\mu g \cdot L^{-1}$) (Kim et al., 2019). Especially in the United States, Vietnam, India, China and parts of Bangladesh (Huang, 2018; Kim et al., 2019), Arsenic exposure in potable water is a major environmental problem (Bandpei et al., 2016; Tiwari et al., 2017). Inorganic arsenate As(V) mainly exists as anionic species in an aerobic environment such as seawater, lakes, and rivers, whereas arsenite As(III) is the main species in anaerobic conditions typical of groundwater and submerged soil (Shi et al., 2019; Zhao et al., 2009). Compared with As(V), As(III) is more mobile, more soluble, and more toxic (Bandpei et al., 2016; Tiwari et al., 2017). Due to the advantages of good scalability, low cost and simple operation, ZVI is identified as an excellent reaction material for sequestration of As(III) and As(V) (Litter et al., 2010). However, in most cases, the traditional ZVI system is inefficient in removing arsenic due to the slow corrosion rate of ZVI (Xue et al., 2013).

Currently, the WMF was adopted to accelerate As(III) and As(V) sequestration by ZVI. It revealed that WMF significantly improved arsenic removal from real As-bearing groundwater by ZVI (Sun et al., 2014). As(III)/As(V) removal was improved by WMF through increasing ZVI corrosion. Sun et al (2017) also studied the role of WMF coupling with ZVI on As(III) sequestration under ClO_4^-, NO_3^-, Cl^-, SO_4^{2-}, $HSiO_3^-$, HCO_3^-, and $H_2PO_4^-$ anion conditions (Sun et al., 2017b). It showed that the anions were very important to stimulate or maintain the role of WMF. In alone ZVI system, NO_3^- and ClO_4^- played little effect on As(III) sequestration, however, SO_4^{2-} and Cl^- could improve As(III) sequestration. In addition, WMF had a minor effect on As(III) sequestration by ZVI in ultrapure water. In the WMF-ZVI system, the anion-enhancing effects were as follows: $SO_4^{2-} > Cl^- > NO_3^- \approx ClO_4^-$. However, $HSiO_3^-$, $H_2PO_4^-$ and HCO_3^- could inhibit ZVI corrosion and the inhibitory effect could be alleviated by the combined effect of SO_4^{2-} and WMF.

The pre-magnetized ZVI technology was adopted to treat As(III) contaminated wastewater (Li et al., 2015). The results showed that pre-magnetized ZVI could keep a longer reactivity than pristine ZVI for the sequestration of As(III). Besides, the pre-magnetization promoted the oxidation of As(III) and the transformation of iron (hydr) oxides from ZVI, and the original corrosion products γ-Fe_2O_3 and Fe_3O_4 were changed to γ-FeOOH, which helped the arsenic sequestration (Figure 7.6). The enhanced effect of pre-magnetized ZVI

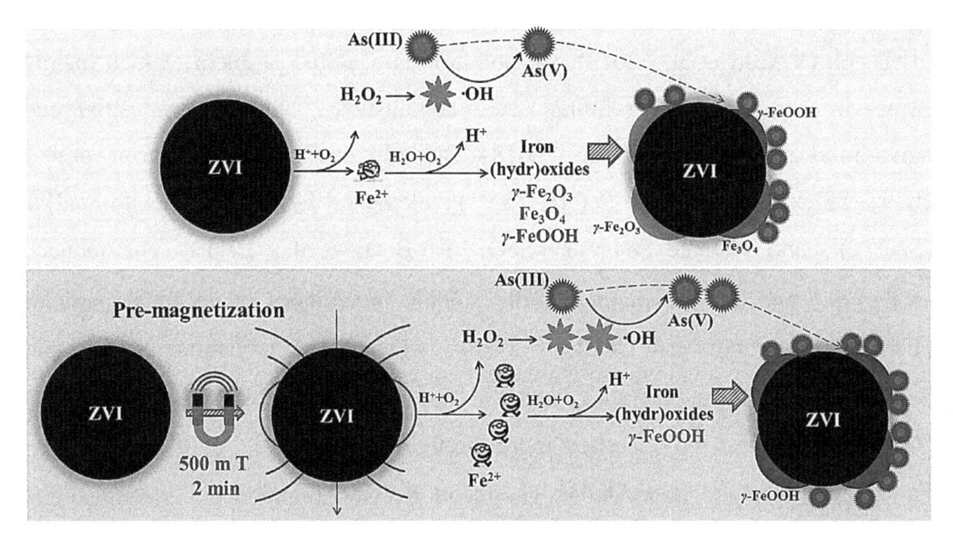

Figure 7.6 Proposed mechanism of pre-magnetization (magnetic memory) promoting the oxidation of As(III) and the transformation of iron (hydr) oxides from ZVI (Pires et al., 2012).

was mainly from the magnetic memory of ZVI.

Taken together with the research, the application of pre-magnetized ZVI and WMF were useful approaches to keep the long-time reactivity of ZVI and presented more excellent performance than pristine ZVI for the sequestration of As(III). It can alleviate the inhibition of $HSiO_3^-$, $H_2PO_4^-$ and HCO_3^- on ZVI corrosion and enhance the transformation of iron (hydr) oxides.

7.2.6 Selenium

Selenium (Se) is a non-metal and essential trace element in the growth of animals, plants, and humans (Constantino et al., 2017; Vinceti et al., 2014), since it has a critical physiological effect on the antioxidant defense systems, immune function, formation of enzymes, and thyroid hormone metabolism (Gong et al., 2018; Pezzarossa et al., 2012). However, high concentrations of Se are considered toxic (Constantino et al., 2017). Se, with its paradoxical biologic characteristics, is a particularly challenging case. The healthy level between the acute toxic level of Se (> 400 μg · day^{-1}) and dietary deficiency of Se (< 40 μg · day^{-1}) are rather narrow (Gore et al., 2010; Ouyang et al., 2018; Vogel et al., 2018). Less than 10 ppb of Se in drinking water is recommended by the WHO and the EU (Zelmanov and Semiat, 2013), while the USEPA limits it to 50 ppb (Vinceti et al., 2013). Se pollution is a global problem, which mainly comes from mining, oil refining, coal combustion, and other industrial and agricultural activities (He et al., 2018; Olegario et al., 2009; Sharma et al., 2019). Se exists as selenide Se(-II) (Se^{2-}), elemental Se (Se^0), selenite Se(IV) (SeO_3^{2-}), and selenate Se(VI) (SeO_4^{2-}) (Bajaj et al., 2011). The reduced species Se^{2-} and Se^0 are mainly hardly soluble precipitates or colloidal particles (Fernández-Martínez and Charlet, 2009), whereas the oxyanions SeO_4^{2-} and SeO_3^{2-} are mobile and soluble, and thus are potentially toxic (Hayashi et al., 2009). Se(IV) can be adsorbed more strongly than Se(VI) onto hematite (α-Fe_2O_3) and goethite (α-FeOOH) (Liang et al., 2015). Se (VI), however, is difficult to be absorbed on various minerals. Therefore, Se(VI) is the most mobile species of Se, which could not easily be sequestrated by traditional adsorption, coagulation, and softening methods (Zhang et al., 2010). Compared to the above traditional methods, the WMF-ZVI system can rapidly reduce Se (VI) to more immobile species, Se(-II), Se(IV), or Se(0), and reductive removal of Se (VI) in this system is more favored.

The performance of the WMF-ZVI system for Se (IV) and Se (VI) sequestration was examined. Liang et al. studied the Se(IV) removal kinetics under WMF with ZVI condition (Liang et al., 2014b). It revealed that WMF induced more excellent enhancement for Se(IV) sequestration in the ZVI system under lower concentration Se(IV) conditions. Moreover, the working pH range of ZVI was significantly extended by the WMF. Compared to the control without WMF, a more rapid Fe^{2+} release and a more dramatic ORP drop occurred in the WMF-ZVI system. The WMF accelerated ZVI corrosion and facilitated the transformation of amorphous iron (hdyr) oxides to γ-FeOOH. The F_L and $F_{\Delta B}$ were speculated to be the main driving force for improving the Se(IV) sequestration in the ZVI system. The effects of WMF and aging on Se(IV) sequestration in the ZVI system were also studied (Liang et al., 2014a). It was shown that the Se(IV) sequestration by ZVI was significantly affected by the combined WMF and aging. It had nearly the same rates of Se(IV) sequestration between 6 and 60 hours in ZVI (aged at pH 6.0) with the WMF system. However, compared with the alone ZVI system, the rate of Se(IV) sequestration in the WMF-ZVI system increased by 1000% to 10000%. In addition, the adsorption-reduction removal mechanism of Se(IV) was changed to direct Se(IV) reduction by WMF in the aged ZVI system.

The effect of the WMF on the depassivation of aged ZVI and Se(IV) removal was studied (Xu et al., 2016b). The results showed that compared to pristine ZVI, the aged ZVI without WMF had much lower reactivity for sequestration of Se(IV). However, the reactivity of all aged ZVI was significantly improved under the WMF condition. In addition, it showed that the intensity of the WMF required for depassivation of aged ZVI had a close relationship with the type of passive film. The magnetic field of 1 mT was strong enough to recover the reactivity of aged ZVI with Fe_3O_4 passive film or a thin α-Fe_2O_3 and γ-FeOOH passive film. However, the intensity of magnetic flux needs to increase to 2 mT or even 5 mT if much thicker and denser passive films or more α-Fe_2O_3 and γ-FeOOH were contained in the aged ZVI.

The rejuvenation effect of WMF on eight types of coexisting solutes (HCO_3^-, NO_3^-, SO_4^{2-}, Cl^-, PO_4^{3-}, SiO_3^{2-}) aged ZVI reactivity toward Se(IV) removal was also studied (Zhang et al., 2018). It was shown that aging significantly influences the application of ZVI for wastewater remediation. The rate of Se(IV) removal by aged ZVI in different backgrounds was ranked as follows: HCO_3^- > NO_3^- > SO_4^{2-} > Cl^- > none (pristine ZVI) > PO_4^{3-} > H_2O > SiO_3^{2-} > synthetic groundwater. Moreover, the Se(IV) removal rates in the aged ZVI system were improved by 430% to 3 960% under the WMF condition. WMF-ZVI also was adopted to facilitate Se(VI) sequestration. Liang et al. studied the performance of the WMF-ZVI system for Se(VI) sequestration (Liang et al., 2015). Compared to less than 4% removal of Se(VI) in the ZVI system in 72 h, 100% of Se(VI) was removed in the WMF-ZVI system in 1.5 h. ZVI was reportedly converted into amorphous Fe_3O_4 and eventually into lepidocrocite (γ-FeOOH). Adsorption of Se(VI) was little, Se(VI) was adsorbed to the ZVI surface then quickly transformed to Se(IV) and then converted to Se(0) (Liang et al., 2015).

Taken together with the studies, WMF is an approach with considerable potential for depassivation of aged ZVI and improving the removal of Se(IV) and Se(VI) from groundwater and wastewater. It can recover and significantly improve the reactivity of eight types of coexisting solutes (HCO_3^-, NO_3^-, SO_4^{2-}, Cl^-, PO_4^{3-}, SiO_3^{2-}) aged ZVI, accelerate ZVI corrosion and facilitate the transformation of amorphous iron (hdyr) oxides to γ-FeOOH.

7.2.7 Antimony (Sb)

Antimony (Sb) is the 9th mined metalloid and it is extensively applied in the production of pigments, automobile brake linings, tracer bullets, small arms, batteries, lead hardener, flameproof retardants, diodes, and semiconductors (Fan et al., 2018; Wilson et al., 2010). Due to Sb mining, spent ammunition, waste incineration, smelting activities, soil runoff, and rock weathering (He, 2007; Wang et al., 2011; Xi et al., 2013), extensive Sb is released into the surface water, groundwater, and soil, causing severe environment pollution (He et al.,

2012; Okkenhaug et al., 2012; Westerhoff et al., 2008). Sb and its compounds have acute toxicity and have been identified as potentially carcinogenic elements (Amarasiriwardena and Wu, 2011; Anjum and Datta, 2012; Oorts et al., 2008). Prolonged exposure to Sb may cause chronic symptoms of sickness (Sundar and Chakravarty, 2010). Thus, Sb and its compounds are listed as persistent organic pollutants (POPs) by the EU (Du et al., 2014) and the USEPA (USEPA, 2009). The WHO, China, and Japan set the safe level of drinking water for Sb at 20 $\mu g \cdot L^{-1}$, 5 $\mu g \cdot L^{-1}$, and 2 $\mu g \cdot L^{-1}$, respectively (China, 2006; WHO, 2011; Zheng et al., 2000).

Sb(V) and Sb(III) are the most common species of Sb (Filella et al., 2002), and Sb(III) is ten times more toxic than Sb(V) (Nam et al., 2009; Oorts et al., 2008). Sb(III) is mainly in the form of $Sb(OH)_3$ in the anoxic environment, while Sb(V) is mainly in the form of $Sb(OH)_6^-$ in the aerobic environment (Li et al., 2015). Sb(III) tends to cooperate with phenolic groups and carboxylic acids (Filella and Williams, 2012; Tella and Pokrovski, 2009). The mobility and solubility of Sb(III) were significantly reduced and the risk of Sb polluting the environment was increased by complexing with these organic ligands (Biver and Shotyk, 2012; Hu and He, 2017). The adsorption (Vithanage et al., 2013; Xi and He, 2013), reverse osmosis, ion exchange (Miao et al., 2014), coagulation-precipitation (Guo et al., 2009), electro-deposition (Bergmann and Koparal, 2011), and bio-sorption (Wu et al., 2012) were used to remove Sb. However, these technologies show some limitations in long-term applications. Such as, reverse osmosis is costly or time-consuming (Kang et al., 2001), and adsorbents are difficult to regenerate (Leng et al., 2012).

The WMF could avoid the above problems and markedly improve ZVI reactivity. The performance of WMF-ZVI on Sb(III) sequestration was explored (Xu et al., 2016b). It indicated that the removal of Sb(III) roughly increased by 600% to 800% under the WMF condition. The WMF delayed the passivation of the ZVI and stimulated its corrosion. The Sb(III) was oxidized into Sb(V) and co-precipitated and adsorbed onto the iron oxides. In addition, the researchers investigated the effect of ZVI dosage, Sb dosage, and initial Sb(III) concentration on Sb(III) removal. The most significant influencing factor was the WMF, which further confirmed the WMF coupling with ZVI for remediation of Sb(III) contaminated groundwater or wastewater can achieve excellent results. Furthermore, the effect of WMF and tartrate on the removal of Sb(III) in the ZVI system was investigated (Fan et al., 2018). The results revealed that the removal of Sb_{tot}(total Sb) and Sb(III) were improved by the WMF. The enhancing effect came from the WMF accelerated ZVI corrosion. Moreover, the results demonstrated that the tartrate significantly decreased Sb removal in the ZVI system. The ZVI corrosion was inhibited by tartrate.

It was reported that ZVI also had a good curing effect on Sb(V) (Li et al., 2015), ZVI was passivated to some extent with the increase of Sb(V) concentration, and the addition of WMF inhibited ZVI passivation, promoted ZVI corrosion, and increased the Sb(V) removal (Li et al., 2015). The removal capacity was improved from 18.1 mg to 39.2 mg Sb(V)/g Fe and the removal rate of Sb(V) was improved by 556% to 771% by WMF. In addition, the results showed that the co-existing humic acid, SiO_3^{2-}, Cl^-, CO_3^{2-}, NO_3^-, and SO_4^{2-} had little effect on Sb(V) sequestration in WMF-ZVI system (Li et al., 2015).

In summary, WMF-ZVI technology is effective for Sb(III) and Sb(V) sequestration and immobilization. The WMF can delay passivation of the ZVI, stimulate ZVI corrosion and eliminate the effect of co-existing humic acid, SiO_3^{2-}, Cl^-, CO_3^{2-}, NO_3^-, and SO_4^{2-} on Sb(V) removal.

7.2.8 Other contaminants and applications

The WMF coupling with ZVI is also used to enhance the phenol degradation (Wang et al., 2015), partial nitrification process (Wang et al., 2017), CH_4 production (Huang et al., 2019), SO_2 removal (Jiang et al., 2008), tributyl phosphate degradation (Ambashta et al., 2011) and emerging contaminants (antibiotic sulfamethazine (SMT), oxytetracycline (OTC), tetracycline (TC), and sulfadiazine (SD)) removal (Pan et al., 2019a). A new magnetic nZVI encapsulated in carbon spheres was utilized to degrade phenol (Wang et al., 2015). The pre-magnetized ZVI/PS process also was used to degrade phenol, the removal rate of phenol was improved by 343% (Table 7.2). Moreover, the potential effect of the external magnetic field on the partial nitrification process was investigated (Wang et al., 2017). It was found that the magnetic field has an excellent performance in promoting the activities of aerobic ammonium oxidizing bacteria of partial nitrification consortium. The 5 mT WMF had the best effect on improving the gene expressions of cell motility and signal transduction. The effect of WMF on CH_4 production in a digester with ZVI was studied (Huang et al., 2019).

It was reported that WMF created a much better reductive environment for reducing propionic acid fermentation, providing additional H_2 for hydrogenotrophic methanogenesis and significantly promoting ZVI corrosion. Compared with the system without ZVI or WMF, the CH_4 production was increased by 77.0% and 124.5% in the WMF-ZVI system. The cast-iron scraps were added in a direct current (DC) magnetically fixed bed to remove SO_2(Jiang et al., 2008). It was found that SO_2 was effectively removed by the corrosion of cast-iron scraps. The surface morphologies of scrap were much looser in the DC magnetic field. The SO_2 removal efficiency was improved by 60% (Table 7.2) and corrosion resistance was reduced. Furthermore, a static magnetic field with nanopowders of iron-nickel and iron was used to degrade tributyl phosphate (Ambashta et al., 2011). It was shown that the external DC electromagnetic field with an intensity of 0.7 T promoted the mineralization of tributyl phosphate in the ZVI Fenton-like system (Ambashta et al., 2011).

The Fenton (ZVI/H_2O_2) process is also increasingly applied for the remediation of wastewater; however, the efficiency is always inhibited when treating salty wastewater. To overcome the problem, the pre-magnetized ZVI/H_2O_2 process was developed. A significant removal enhancement was obtained for treating the salty wastewater by this technology (Pan et al., 2018b). Orange G was the target pollutant, and Na_2SO_4 was used as the model salty system. Compared to the ZVI/H_2O_2 process, the degradation rates of Orange G in the pre-magnetized ZVI/H_2O_2 process were increased by 120% to 17 120%. In addition, the H_2O_2 dosage saved 75% in the salty reaction system. The enhancement came from the increased ZVI corrosion, resulting in the rapid H_2O_2 catalysis and the formation of hydroxyl radical (· OH). The pre-magnetized ZVI/H_2O_2 process enhanced NaCl, $NaNO_3$ and Na_2HPO_4 salty systems by 590%, 2 640% and 9 370%, respectively. In addition, the results of the pre-magnetized ZVI/H_2O_2 process for treating the real petrochemical wastewater and reverse osmosis concentrated wastewater showed that the process is cost-effective and the COD removal efficiency improved by 94.4% and 59.9%, respectively (Table 7.2) (Pan et al., 2018b). The pre-magnetized ZVI/H_2O_2 Fenton-like system also showed excellent performance in treating citric acid wastewater, which was the real wastewater from a food processing company (Huang et al., 2018). The H_2O_2 dosage was reduced by 27.5% and chemical oxygen demand (COD) removal increased by 38.4% in this new system (Table 7.2).

Recently, the trace organic contaminants in pharmaceuticals and personal care products (PPCPs) also aroused wide public concern. They are frequently detected in wastewater treatment plants, rivers and groundwater (Yang et al., 2017; Zhang et al., 2018). The concentration of PPCPs is ranging from $ng \cdot L^{-1}$ to $\mu g \cdot L^{-1}$, however, a long-term negative impact on human health is presented because of their persistence and variety of properties (Yang et al., 2017). The study revealed that the amorphous ZVI microspheres significantly improved the reduction rate constant of thiamphenicol (Shen et al., 2020). In addition, the nZVI and iron powder showed a positive effect on the production of methane and the removal of chlorinated PPCPs (clofibric acid and triclosan). However, they did not show significant improvement in the removal of other PPCPs (eg. losartan, propyl paraben and ketoprofen) (Suanon et al., 2017). Here, the ZVI/persulfate and ZVI/H_2O_2 Fenton-like processes were adopted to extend the applicability for the removal of PPCPs (Guo et al., 2020; Minella et al., 2019; Wu et al., 2020).

The previous studies about the degradation of nine typical PPCPs (SMT, TC, sulfamethoxazole phenacetin, carbamazepine, paracetamol, norfloxacin, indomethacin, and bisphenol A) by the nZVI/persulfate Fenton-like process showed that the removal efficiency of PPCPs ranged from 77% to 100%, suggesting that the nZVI/persulfate Fenton-like process had excellent effectiveness and wide applicability for remediation of PPCPs contaminated (waste) water (Wu et al., 2020). The removal of sulfadiazine was significantly improved by the sulfidated ZVI-activated persulfate process (S-ZVI/persulfate Fenton-like process) (Guo et al., 2020) and the ibuprofen was effectively removed from the synthetic and actual wastewater by the ZVI/H_2O_2 Fenton-like process (Minella et al., 2019). Based on the above studies and the pre-magnetized ZVI/H_2O_2 Fenton-like system, the combination of ultraviolet (UV) and pre-magnetized ZVI/H_2O_2 Fenton-like system was further used for remediation of the real wastewater from secondary wastewater effluents contaminated with PPCPs (SMT) (Pan et al., 2019a). The results showed that this method could eliminate SMT within 30 minutes and improve the SMT removal rate by 180%. At the same time, the mineralization capacity of this process was outstanding, the removal of total organic carbon (TOC) reached 92.1%, while the mineralization capacity of UV/ZVI/H_2O_2 and UV/H_2O_2 systems were only 72.1% and 53.9%, respectively (Table 7.2).

Furthermore, based on the above studies and the pre-magnetized ZVI/persulfate system, the combination of UV and pre-magnetized ZVI/persulfate systems (Figure 7.7) was developed to dispose of the real wastewater contaminated with PPCPs (SMT, OTC, TC, and SD) (Pan et al., 2019b). It was shown that OTC, TC, and SMT were completely removed in 30 minutes, 98.4% SD was removed in 60 minutes, while ZVI/PS, pre-magnetized ZVI/PS and UV processes only removed less than 10%, 20% and 60% of emerging pollutants in 60 minutes. EPR spectra and scavenging tests confirmed that not only $SO_4^-\cdot$ and $\cdot OH$ radicals generated faster and more (Figure 7.7), but also their contribution to SMT removal was greater, which improved the SMT removal rate by 226% (Table 7.2).

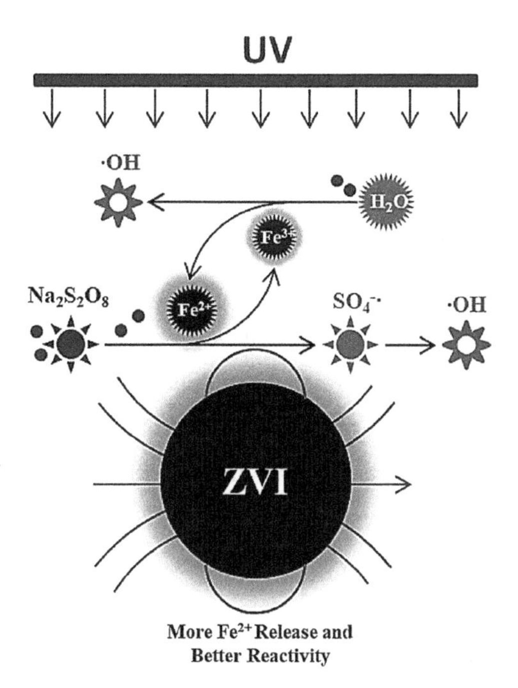

Figure 7.7 Promotion mechanism of the UV coupling with pre-magnetized ZVI/persulfate system for removal of emerging contaminants (SMT, OTC, TC, and SD) (Pan et al., 2019b).

Moreover, the combination of ultrasound (US) and pre-magnetized ZVI/persulfate systems was used to dispose of the SMT (Pan et al., 2018a). Compared with the US/ZVI/PS system, the k_{obs} of SMT in the US/pre-magnetized ZVI/persulfate system was increased by 330% and the synergy factor was increased by 125%. The results revealed that $\cdot OH$ and $SO_4^- \cdot$ were produced faster and more in the US/pre-magnetized ZVI/persulfate system, and it showed a much better synergy between WMF and US than the US/ZVI/PS system.

Based on the views above, the addition of reduction assistants such as WMF, WMF+UV, WMF+US, WMF+H_2O_2, and WMF+PS, is very effective strategy to overcome the above-mentioned disadvantages of ZVI and improve contaminants removal.

7.3 Conclusions and suggestions

The application of WMF coupling with ZVI and magnetic ZVI derivative technologies for sustainable groundwater and wastewater treatment has made certain progress since it was invented. This review shows that even though these technologies are relatively young and developing, a substantial amount of work has been done, to improve WMF-ZVI performance, leading to the generation of WMF combined with AOPs technology, such as the WMF coupling with ZVI/H_2O_2 Fenton-like system, pre-magnetized ZVI/H_2O_2 Fenton-like system, WMF coupling with ZVI/persulfate system and pre-magnetized ZVI/persulfate system, and slowly transforming it from innovative technology to advanced technology. The spectrum of contaminants that could be treated with WMF coupling with ZVI has been broadened, owing to the discovery of WMF coupling with ZVI/H_2O_2 Fenton-like system or WMF coupling with ZVI/persulfate system and understanding of the removal mechanisms.

The efficiency of these processes could be sustained under the WMF condition or by a recovery of magnetism of pre-magnetized ZVI. In addition, the AOPs based on pre-magnetized ZVI or ZVI under WMF condition is a more promising technology instead of AOPs since it does not need sophisticated equipment, costly materials, or an additional energy source. It is more effective for contaminant removal and does not change the present treatment process of groundwater and wastewater. Based on the reasons listed above, it is reasonable to assume that WMF coupling with the ZVI/AOPs system is the future development direction of this cutting-edge technology.

The application of WMF coupling with ZVI and magnetic ZVI derivative technologies for the remediation of contaminated groundwater and wastewater has gained extensive attention. Numerous experiments have proved it has excellent remediation efficiency, while problems and shortages of that are indispensable. Some suggestions on the development and application of this technology are put forward.

a. The reaction of various contaminants in the pre-magnetized ZVI or WMF-ZVI system is more complex than expected. This research has thrown up many questions in need of further investigation. More detailed work should be performed to reveal the removal mechanism of contaminants by pre-magnetized ZVI or ZVI under the WMF condition.

b. Most of the studies on pre-magnetized ZVI or WMF-ZVI technology refer to the optimization of experimental conditions for developing a more effective technique. Pre-magnetized ZVI or WMF-ZVI technology is useful in the remediation of groundwater and wastewater, however, the technology still has poor performance on some specific contaminants. To further improve the performance of technology, the pre-magnetized ZVI or ZVI with other oxidants, reducing agents, or other AOPs technologies under the WMF condition should be further explored and developed.

c. Most studies on the coupling of ZVI and WMF are performed on the laboratory scale, and the results show that this technology has high performance for contaminant removal. Considering the simultaneous presence of multiple organic matters, inorganic minerals and pollutants in groundwater and wastewater, the effect of various inorganic minerals and organic matters on the removal of multiple pollutants and ZVI activity under the WMF condition need to be studied urgently. In addition, the laboratory conditions are far from the real field environment. Therefore, larger-scale field remediation tests are urgently needed to explore the commercial application of ZVI under the WMF conditions.

d. Construction and implementation of a magnetic field in large-scale remediation sites are not convenient. Therefore, it is necessary to study and optimize appropriate methods to ensure the standard operation of magnetic field equipment.

Appendix
Abbreviation List

Abbreviation	English Full Name	Abbreviation	English Full Name
WMF	weak magnetic field	2,4-D	2,4-dichlorophenoxyacetic acid
ZVI	zero-valent iron	CB	chlorobenzene
nZVI	nanosized ZVI	TCAM	trichloroacetamide
pre-ZVI	pre-magnetized ZVI	2,4,6-TCP	2,4,6-trichlorophenol
NE-nZVI	non-electrosprayed nanometer ZVI	2,4-DCP	2,4-dichlorophenol
E-nZVI	electrosprayed nanometer ZVI	4-CP	4-chlorophenol
Fe_3O_4	magnetite	TCE	trichloroethylene
α-Fe_2O_3	hematite	PCE	tetrachloroethylene
γ-Fe_2O_3	maghemite	1,1,1-TCA	1,1,1-trichloroethane
α-FeOOH	goethite	PCB	polychlorinated biphenyls
γ-FeOOH	lepidocrocite	DDT	1,1,1-trichloro-2,2-bis (4-chlorophenyl) ethane
$FeCO_3$	siderite	NPs	nitrophenols
$CaCO_3$	calcite	PNP	4-nitrophenol (*p*-nitrophenol)
WHO	World Health Organization	EDTA	Ethylene Diamine Tetraacetic Acid
ORP	oxidation-reduction potential	OG	Orange G
H_2O_2	hydrogen peroxide	AR27	amaranth
PS	persulfate	RhB	Rhodamine B
$SO_4^{-}\cdot$	sulfate radicals	SMT	sulfamethazine
$\cdot OH$	hydroxyl radicals	OTC	oxytetracycline
$F_{\Delta B}$	magnetic gradient force	TC	tetracycline
F_L	Lorentz force	SD	sulfadiazine
AOPs	Advanced Oxidation Processes	DC	direct current
k_{obs}	first-order rate constants	UV	ultraviolet
COCs	chlorinated organic compounds	COD	chemical oxygen demand
POPs	persistent organic pollutants	AC	alternating current
PPCPs	pharmaceuticals and personal care products	US	ultrasound

References

- Abd-Elaty, I., Pugliese, L., Zelenakova, M., Mesaros, P., Shinawi, A. E., 2020. Simulation-based solutions reducing soil and groundwater contamination from fertilizers in arid and semi-arid regions: case study the eastern Nile Delta, Egypt. *International Journal of Environmental Research and Public Health* 17(24), 9373–9391. https://doi.org/10.3390/ijerph17249373.
- Aihemaiti, A., Gao, Y., Meng, Y., Chen, X., Liu, J., Xiang, H., Xu, Y., Jiang, J., 2020. Review of plant-vanadium physiological interactions, bioaccumulation, and bioremediation of vanadium-contaminated sites. *Science of the Total Environment* 712, 135637. https://doi.org/10.1016/j.scitotenv.2019.135637.
- Al-Abduly, A. J., Basfar, A. A., Algamdi, M. S., Alsohaimi, I. H., Alharbi, A. A., Alkhedhair, A.M., 2021. Remediation and detoxification of water samples contaminated with 2, 4, 6-trichlorophenol by gamma radiation and ozonation. *Radiation Physics and Chemistry* 184, 109423. https://doi.org/10.1016/j.radphyschem.2021.109423.
- Altowayti, W. A. H., Othman, N., Shahir, S., Alshalif, A. F., Al-Gheethi, A. A., Al-Towayti, F.A.H., Saleh, Z.M., Haris, S.A., 2021. Removal of arsenic from wastewater by using different technologies and adsorbents: a review. International Journal of *Environmental Science and Technology* 19, 9243–9266. https://doi.org/10.1007/s13762-021-03660-0.
- Amano, H., Nakagawa, K., Berndtsson, R., 2016. Groundwater geochemistry of a nitrate-contaminated agricultural site. *Environmental Earth Sciences* 75(15), 1145–1159. https://doi.org/10.1007/s12665-016-5968-8.
- Amarasiriwardena, D., Wu, F., 2011. Antimony: Emerging toxic contaminant in the environment. *Microchemical Journal* 97(1), 1–3. https://doi.org/10.1016/j.microc.2010.07.009.
- Ambashta, R.D., Repo, E., Sillanpää, M., 2011. Degradation of tributyl phosphate using nanopowders of iron and iron-nickel under the influence of a static magnetic field. *Industrial & Engineering Chemistry Research* 50(21), 11771–11777. https://doi.org/10.1021/ie102121e.
- Amoako-Nimako, G. K., Yang, X., Chen, F., 2021. Denitrification using permeable reactive barriers with organic substrate or zero-valent iron fillers: controlling mechanisms, challenges, and future perspectives. *Environmental Science and Pollution Research* 28(17),

21045–21064. https://doi.org/10.1007/s11356-021-13260-7.

- An, T., Zu, L., Li, G., Wan, S., Mai, B., Wong, P.K., 2011. One-step process for debromination and aerobic mineralization of tetrabromobisphenol-A by a novel Ochrobactrum sp. T isolated from an e-waste recycling site. *Bioresource Technology* 102(19), 9148–9154. https://doi.org/10.1016/j.biortech.2011.06.080.
- Anandan, S., Kumar Ponnusamy, V., Ashokkumar, M., 2020. A review on hybrid techniques for the degradation of organic pollutants in aqueous environment. *Ultrasonics Sonochemistry* 67, 105130. https://doi.org/10.1016/j.ultsonch.2020.105130.
- Andrade, D.C., dos Santos, E.V., 2020. Combination of electrokinetic remediation with permeable reactive barriers to remove organic compounds from soils. Current Opinion in Electrochemistry 22, 136–144. https://doi.org/10.1016/j.coelec.2020.06.002.
- Angai, J.U., Ptacek, C.J., Pakostova, E., Bain, J.G., Verbuyst, B.R., Blowes, D.W., 2022. Removal of arsenic and metals from groundwater impacted by mine waste using zero-valent iron and organic carbon: Laboratory column experiments. *Journal of Hazardous Materials* 424, 127295. https://doi.org/10.1016/j.jhazmat.2021.127295.
- Anjum, A., Datta, M., 2012. Adsorptive removal of antimony (III) using modified montmorillonite: A study on sorption kinetics. *Journal of Analytical Sciences, Methods and Instrumentation* 02(03), 167–175. https://doi.org/10.4236/jasmi.2012.23027.
- Ansaf, K.V.K., Ambika, S., Nambi, I.M., 2016. Performance enhancement of zero valent iron based systems using depassivators: optimization and kinetic mechanisms. *Water Research* 102, 436–444. https://doi.org/10.1016/j.watres.2016.06.064.
- Arjoon, A., Olaniran, A.O., Pillay, B., 2012. Co-contamination of water with chlorinated hydrocarbons and heavy metals: Challenges and current bioremediation strategies. *International Journal of Environmental Science and Technology* 10(2), 395–412. https://doi.org/10.1007/s13762-012-0122-y.
- Arora, P.K., 2012. Metabolism of para-nitrophenol in Arthrobacter sp. SPG. *Environmental Research Management* 3, 52–57.
- Arora, P.K., Sasikala, C., Ramana Ch, V., 2012. Degradation of chlorinated nitroaromatic compounds. *Applied Microbiology and Biotechnology* 93(6), 2265–2277. https://doi.org/10.1007/s00253-012-3927-1.
- Arora, P.K., Srivastava, A., Singh, V.P., 2014. Bacterial degradation of nitrophenols and their derivatives. *Journal of Hazardous Materials* 266, 42–59. https://doi.org/10.1016/j.jhazmat.2013.12.011.
- Ashad, H., 2023. Minimizing the impact of microorganism intrusion on the concrete physical and mechanical properties with nickel waste as a partial substitution for cement. *Heliyon* 9(2), e13303. https://doi.org/10.1016/j.heliyon.2023.e13303.
- Azubuike, C.C., Chikere, C.B., Okpokwasili, G.C., 2016. Bioremediation techniques-

classification based on site of application: principles, advantages, limitations and prospects. *World Journal of Microbiology and Biotechnology* 32(11), 180–198. https://doi.org/10.1007/s11274-016-2137-x.

- Bae, S., Joo, J.B., Lee, W., 2017. Reductive dechlorination of carbon tetrachloride by bioreduction of nontronite. *Journal of Hazardous Materials* 334, 104–111. https://doi.org/https://doi.org/10.1016/j.jhazmat.2017.03.066.
- Bae, S., Lee, W., 2012. Enhanced reductive degradation of carbon tetrachloride by biogenic vivianite and Fe(II). *Geochimica et Cosmochimica Acta* 85, 170–186. https://doi.org/https://doi.org/10.1016/j.gca.2012.02.023.
- Bafana, A., Devi, S.S., Chakrabarti, T., 2011. Azo dyes: past, present and the future. *Environmental Research* 19, 350–371.
- Bahr, J.M., Rubin, J., 1987. Direct comparison of kinetic and local equilibrium formulations for solute transport affected by surface reactions. *Water Resources Research* 23, 438–452.
- Bajaj, M., Eiche, E., Neumann, T., Winter, J., Gallert, C., 2011. Hazardous concentrations of selenium in soil and groundwater in North-West India. *Journal of Hazardous Materials* 189(3), 640–646. https://doi.org/10.1016/j.jhazmat.2011.01.086.
- Bandpei, A.M., Mohseni, S.M., Sheikhmohammadi, A., Sardar, M., Sarkhosh, M., Almasian, M., Avazpour, M., Mosallanejad, Z., Atafar, Z., Nazari, S., Rezaei, S., 2016. Optimization of arsenite removal by adsorption onto organically modified montmorillonite clay: experimental & theoretical approaches. *Korean Journal of Chemical Engineering* 34(2), 376–383. https://doi.org/10.1007/s11814-016-0287-z.
- Bekele, D.N., Du, J., de Freitas, L.G., Mallavarapu, M., Chadalavada, S., Naidu, R., 2019. Actively facilitated permeable reactive barrier for remediation of TCE from a low permeability aquifer: field application. *Journal of Hydrology* 572, 592–602. https://doi.org/10.1016/j.jhydrol.2019.03.059.
- Bergmann, M.E., Koparal, A.S., 2011. Electrochemical antimony removal from accumulator acid: results from removal trials in laboratory cells. *Journal of Hazardous Materials* 196, 59–65. https://doi.org/10.1016/j.jhazmat.2011.08.073.
- Bertolini, M., Zecchin, S., Beretta, G.P., De Nisi, P., Ferrari, L., Cavalca, L., 2021. Effectiveness of permeable reactive bio-barriers for bioremediation of an organohalide-polluted aquifer by natural-occurring microbial community. *Water* 13(17), 2442–2464. https://doi.org/10.3390/w13172442.
- Bilardi, S., Ielo, D., Moraci, N., Calabrò, P.S., 2016. Reactive and hydraulic behavior of permeable reactive barriers constituted by Fe^0 and granular mixtures of Fe^0/pumice. *Procedia Engineering* 158, 446–451. https://doi.org/10.1016/j.proeng.2016.08.470.
- Birch, G.F., Drage, D.S., Thompson, K., Eaglesham, G., Mueller, J.F., 2015. Emerging contaminants (pharmaceuticals, personal care products, a food additive and pesticides) in

waters of Sydney estuary, Australia. *Marine Pollution Bulletin* 97(1-2), 56-66. https://doi.org/10.1016/j.marpolbul.2015.06.038.

- Biver, M., Shotyk, W., 2012. Experimental study of the kinetics of ligand-promoted dissolution of stibnite (Sb_2S_3). *Chemical Geology* 294-295, 165-172. https://doi.org/10.1016/j.chemgeo.2011.11.009.
- Breitenstein, A., Andreesen, J.R., Lechner, U., 2007. Analysis of an anaerobic chemostat population stably dechlorinating 2,4,6-trichlorophenol. *Engineering in Life Sciences*, 380-387. https://doi.org/10.1002/elsc.200720205.
- Brusseau, M.L., 2020. Simulating PFAS transport influenced by rate-limited multi-process retention. *Water Research* 168, 115179. https://doi.org/10.1016/j.watres.2019.115179.
- Brusseau, M.L., Yan, N., Van Glubt, S., Wang, Y., Chen, W., Lyu, Y., Dungan, B., Carroll, K.C., Holguin, F.O., 2019. Comprehensive retention model for PFAS transport in subsurface systems. *Water Research* 148, 41-50. https://doi.org/10.1016/j.watres.2018.10.035.
- Bulusu, R.K.M., Wandell, R.J., Zhang, Z., Farahani, M., Tang, Y., Locke, B.R., 2020. Degradation of PFOA with a nanosecond-pulsed plasma gas-liquid flowing film reactor. *Plasma Processes and Polymers* 17(8), 2000074. https://doi.org/10.1002/ppap.202000074.
- Campos-Pereira, H., Kleja, D.B., Sjostedt, C., Ahrens, L., Klysubun, W., Gustafsson, J. P., 2020. The adsorption of per- and polyfluoroalkyl substances (PFASs) onto ferrihydrite is governed by surface charge. *Environmental Science & Technology* 54(24), 15722-15730. https://doi.org/10.1021/acs.est.0c01646.
- Cancelo-González, J., Prieto, D.M., Paradelo, R., Barral, M.T., 2015. A microcosm study of permeable reactive barriers filled with granite powder and compost for the treatment of water contaminated with Cr (VI). *Spanish Journal of Soil Science* 5, 180-190. https://doi.org/10.3232/sjss.2015.V5.N2.07.
- Cao, V., Yang, H., Ndé-Tchoupé, A.I., Hu, R., Gwenzi, W., Noubactep, C., 2020. Tracing the scientific history of FeO-based environmental remediation prior to the advent of permeable reactive barriers. *Processes* 8, 977-991. https://doi.org/10.3390/pr8080977.
- Carvalho, M.F., Ferreira Jorge, R., Pacheco, C.C., De Marco, P., Castro, P.M., 2005. Isolation and properties of a pure bacterial strain capable of fluorobenzene degradation as sole carbon and energy source. *Environmental Microbiology* 7(2), 294-298. https://doi.org/10.1111/j.1462-2920.2004.00714.x.
- Chae, J.C., Song, B., Zylstra, G.J., 2008. Identification of genes coding for hydrolytic dehalogenation in the metagenome derived from a denitrifying 4-chlorobenzoate degrading consortium. *Federation of European Microbiological Societies* 281(2), 203-209. https://doi.org/10.1111/j.1574-6968.2008.01106.x.
- Chakraborty, J., Rajput, V., Sapkale, V., Kamble, S., Dharne, M., 2021. Spatio-temporal

resolution of taxonomic and functional microbiome of Lonar soda lake of India reveals metabolic potential for bioremediation. *Chemosphere* 264, 128574. https://doi.org/10.1016/j.chemosphere.2020.128574.

- Chen, G., Liu, H., 2017. Understanding the reduction kinetics of aqueous vanadium(V) and transformation products using rotating ring-disk electrodes. *Enviormental Science & Technology* 51(20), 11643–11651. https://doi.org/10.1021/acs.est.7b02021.
- Chen, H., Cao, Y., Wei, E., Gong, T., Xian, Q., 2016. Facile synthesis of graphene nano zero-valent iron composites and their efficient removal of trichloronitromethane from drinking water. *Chemosphere* 146, 32–39. https://doi.org/10.1016/j.chemosphere.2015.11.095.
- Chen, L., Jin, S., Fallgren, P.H., Swoboda-Colberg, N.G., Liu, F., Colberg, P.J., 2012. Electrochemical depassivation of zero-valent iron for trichloroethene reduction. *Journal of Hazardous Materials* 239–240, 265–269. https://doi.org/10.1016/j.jhazmat.2012.08.074.
- Chen, L., Liu, F., Liu, Y., Dong, H., Colberg, P. J., 2011. Benzene and toluene biodegradation down gradient of a zero-valent iron permeable reactive barrier. *Journal of Hazardous Materials* 188(1–3), 110–115. https://doi.org/10.1016/j.jhazmat.2011.01.076.
- Chen, L., Liu, J., Zhang, W., Zhou, J., Luo, D., Li, Z., 2021. Uranium (U) source, speciation, uptake, toxicity and bioremediation strategies in soil-plant system: a review. *Journal of Hazardous Materials* 413, 125319. https://doi. org/10. 1016/j. jhazmat. 2021. 125319.
- Chen, Q., Wei, J.C., Jia, C.P., Wang, H.M., Shi, L.Q., Liu, S.L., Ning, F.Z., Ji, Y. H., Dong, F. Y., Jia, Z. W., Hao, D. C., 2019. Groundwater selenium level and its enrichment dynamics in seawater intrusion area along the northern coastal zones of Shandong Province, China. *Geochemistry International* 57(11), 1236–1242. https://doi.org/10.1134/s0016702919110065.
- Chen, S., Wang, F., Chu, W., Li, X., Wei, H., Gao, N., 2019. Weak magnetic field accelerates chloroacetamide removal by zero-valent iron in drinking water. *Chemical Engineering Journal* 358, 40–47. https://doi.org/10.1016/j.cej.2018.09.212.
- Chitrakar, R., Tezuka, S., Sonoda, A., Sakane, K., Ooi, K., Hirotsu, T., 2006. Phosphate adsorption on synthetic goethite and akaganeite. *Journal of Colloid and Interface Science* 298(2), 602–608. https://doi.org/https://doi.org/10.1016/j.jcis.2005.12.054.
- Christoforidis, K.C., Serestatidou, E., Louloudi, M., Konstantinou, I.K., Milaeva, E.R., Deligiannakis, Y., 2011. Mechanism of catalytic degradation of 2,4,6-trichlorophenol by a Fe-porphyrin catalyst. *Applied Catalysis B: Environmental* 101(3–4), 417–424. https://doi.org/10.1016/j.apcatb.2010.10.011.
- Chubar, N., Gerda, V., Szlachta, M., Yablokova, G., 2021. Effect of Fe oxidation state (+2 versus +3) in precursor on the structure of Fe oxides/carbonates-based composites examined by XPS, FTIR and EXAFS. *Solid State Sciences* 121, 106752–106765. https://

doi.org/10.1016/j.solidstatesciences.2021.106752.

- Cobas, M., Ferreira, L., Tavares, T., Sanroman, M.A., Pazos, M., 2013. Development of permeable reactive biobarrier for the removal of PAHs by Trichoderma longibrachiatum. *Chemosphere* 91(5), 711–716. https://doi.org/10.1016/j.chemosphere.2013.01.028.
- Constantino, L.V., Quirino, J.N., Monteiro, A.M., Abrao, T., Parreira, P.S., Urbano, A., Santos, M.J., 2017. Sorption-desorption of selenite and selenate on Mg-Al layered double hydroxide in competition with nitrate, sulfate and phosphate. *Chemosphere* 181, 627–634. https://doi.org/10.1016/j.chemosphere.2017.04.071.
- Cousins, I.T., Johansson, J.H., Salter, M.E., Sha, B., Scheringer, M., 2022. Outside the safe operating space of a new planetary boundary for per- and polyfluoroalkyl substances (PFAS). *Environmental Science & Technology* 56(16), 11172–11179. https://doi.org/10.1021/acs.est.2c02765.
- Cui, Y., Ren, H., Yang, C., Yan, X., 2019. Room-temperature synthesis of microporous organic network for efficient adsorption and removal of tetrabromobisphenol A from aqueous solution. *Chemical Engineering Journal* 368, 589–597. https://doi.org/10.1016/j.cej.2019.02.153.
- Culver, T.B., Hallisey, S.P., Sahoo, D., Deitsch, J.J., Smith, J.A., 1997. Modeling the desorption of organic contaminants from long-term contaminated soil using distributed mass transfer rates. *Environmental Science & Technology* 31(6), 1581–1588.
- De Pourcq, K., Ayora, C., Garcia-Gutierrez, M., Missana, T., Carrera, J., 2015. A clay permeable reactive barrier to remove Cs-137 from groundwater: Column experiments. *Journal of Environmental Radioactivity* 149, 36–42. https://doi.org/10.1016/j.jenvrad.2015.06.029.
- Dermatas, D., 2017. Waste management and research and the sustainable development goals: Focus on soil and groundwater pollution. *Waste Management & Research* 35(5), 453–455. https://doi.org/10.1177/0734242X17706474.
- Ding, R., Wu, Y., Yang, F., Xiao, X., Li, Y., Tian, X., Zhao, F., 2021. Degradation of low-concentration perfluorooctanoic acid via a microbial-based synergistic method: assessment of the feasibility and functional microorganisms. *Journal of Hazardous Materials* 416, 125857. https://doi.org/10.1016/j.jhazmat.2021.125857.
- Ding, Y., Zhu, L., Wang, N., Tan, H., 2013. Sulfate radicals induced degradation of tetrabromobisphenol A with nanoscaled magnetic CuFe2O4 as a heterogeneous catalyst of peroxymonosulfate. *Applied Catalysis B: Environmental* 129, 153–162. https://doi.org/10.1016/j.apcatb.2012.09.015.
- Do, S.H., Kwon, Y.J., Kong, S.H., 2011. Feasibility study on an oxidant-injected permeable reactive barrier to treat BTEX contamination: adsorptive and catalytic characteristics of waste-reclaimed adsorbent. *Journal of Hazardous Materials* 191(1–3), 19–25. https://doi.org/10.1016/j.jhazmat.2011.03.115.

- Dong, H., Huang, L., Zhao, L., Zeng, Q., Liu, X., Sheng, Y., Shi, L., Wu, G., Jiang, H., Li, F., Zhang, L., Guo, D., Li, G., Hou, W., Chen, H., 2022. A critical review of mineral-microbe interaction and co-evolution: mechanisms and applications. *National Science Review* 9(10), 128-149. https://doi.org/10.1093/nsr/nwac128.
- Dorathi, P.J., Kandasamy, P., 2012. Dechlorination of chlorophenols by zero valent iron impregnated silica. *Journal of Environmental Sciences* 24(4), 765-773. https://doi.org/10.1016/s1001-0742(11)60817-6.
- Du, J., Che, D., Li, X., Guo, W., Ren, N., 2017. Factors affecting p-nitrophenol removal by microscale zero-valent iron coupling with weak magnetic field. *RSC Advances* 7(30), 18231-18237. https://doi.org/10.1039/C7RA02002C10.1039/c7ra02002c.
- Du, J., Guo, W., Che, D., Ren, N., 2018. Weak magnetic field for enhanced oxidation of sulfamethoxazole by Fe^0/H_2O_2 and Fe^0/persulfate: Performance, mechanisms, and degradation pathways. *Chemical Engineering Journal* 351, 532-539. https://doi.org/10.1016/j.cej.2018.06.094.
- Du, X., Qu, F., Liang, H., Li, K., Yu, H., Bai, L., Li, G., 2014. Removal of antimony (III) from polluted surface water using a hybrid coagulation-flocculation-ultrafiltration (CF-UF) process. *Chemical Engineering Journal* 254, 293-301. https://doi.org/10.1016/j.cej.2014.05.126.
- Economou-Eliopoulos, M., Megremi, I., 2021. Contamination of the soil-groundwater-crop system: environmentalrisk and opportunities. *Minerals* 11(7), 775-792. https://doi.org/10.3390/min11070775.
- Eglal, M.M., Ramamurthy, A.S., 2015. Competitive adsorption and oxidation behavior of heavy metals on nZVI coated with TEOS. *Water Environment Research* 87(11), 2018-2026. https://doi.org/10.2175/106143015X14338845155020.
- Egli, C., Scholtz, R., Cook, A.M., Leisinger, T., 1987. Anaerobic dechlorination of tetrachloromethane and 1, 2-dichloroethane to degradable products by pure cultures of Desulfobacterium sp. and Methanobacterium sp. *Fems Microbiology Letters* 43, 257-261.
- Eichenbaum, G., Johnson, M., Kirkland, D., O'Neill, P., Stellar, S., Bielawne, J., DeWire, R., Areia, D., Bryant, S., Weiner, S., Desai-Krieger, D., Guzzie-Peck, P., Evans, D.C., Tonelli, A., 2009. Assessment of the genotoxic and carcinogenic risks of p-nitrophenol when it is present as an impurity in a drug product. *Regulatory Toxicology and Pharmacology: RTP* 55(1), 33-42. https://doi.org/10.1016/j.yrtph.2009.05.018.
- Eljamal, O., Eljamal, R., Maamoun, I., Khalil, A.M.E., Shubair, T., Falyouna, O., Sugihara, Y., 2022. Efficient treatment of ammonia-nitrogen contaminated waters by nano zero-valent iron/zeolite composite. *Chemosphere* 287(Pt 1), 131990. https://doi.org/10.1016/j.chemosphere.2021.131990.
- Eljamal, O., Maamoun, I., Alkhudhayri, S., Eljamal, R., Falyouna, O., Tanaka, K.,

Kozai, N., Sugihara, Y., 2022. Insights into boron removal from water using Mg-Al-LDH: Reaction parameters optimization & 3D-RSM modeling. *Journal of Water Process Engineering* 46, 102608. https://doi.org/10.1016/j.jwpe.2022.102608.

- Eljamal, O., Sasaki, K., Hirajima, T., 2011. Numerical simulation for reactive solute transport of arsenic in permeable reactive barrier column including zero-valent iron. *Applied Mathematical Modelling* 35(10), 5198-5207. https://doi.org/10.1016/j.apm.2011.04.040.
- Eljamal, O., Thompson, I.P., Maamoun, I., Shubair, T., Eljamal, K., Lueangwattanapong, K., Sugihara, Y., 2020. Investigating the design parameters for a permeable reactive barrier consisting of nanoscale zero-valent iron and bimetallic iron/copper for phosphate removal. *Journal of Molecular Liquids* 299, 112144. https://doi.org/10.1016/j.molliq.2019.112144.
- Engelmann, C., Sookhak Lari, K., Schmidt, L., Werth, C. J., Walther, M., 2021. Towards predicting DNAPL source zone formation to improve plume assessment: Using robust laboratory and numerical experiments to evaluate the relevance of retention curve characteristics. *Journal of Hazards Materials* 407, 124741. https://doi. org/10. 1016/j. jhazmat.2020.124741.
- Evich, M. G., Davis, M. J. B., McCord, J. P., Acrey, B., Awkerman, J. A., Knappe, D.R.U., Lindstrom, A. B., Speth, T. F., Tebes-Stevens, C., Strynar, M. J., Wang, Z., Weber, E. J., Henderson, W. M., Washington, J. W., 2022. Per- and polyfluoroalkyl substances in the environment. *Science* 375 (6580), eabg9065. https://doi. org/10. 1126/science.abg9065.
- Ewlad-Ahmed, A.M., Morris, M., Holmes, J., Belton, D.J., Patwardhan, S.V., Gibson, L.T., 2021. Green nanosilicas for monoaromatic hydrocarbons removal from air. *Silicon* 14(4), 1447-1454. https://doi.org/10.1007/s12633-020-00924-1.
- Falk, S., Stahl, T., Fliedner, A., Rudel, H., Tarricone, K., Brunn, H., Koschorreck, J., 2019. Levels, accumulation patterns and retrospective trends of perfluoroalkyl acids (PFAAs) in terrestrial ecosystems over the last three decades. *Environmental Pollution* 246, 921-931. https://doi.org/10.1016/j.envpol.2018.12.095.
- Falyouna, O., Maamoun, I., Bensaida, K., Tahara, A., Sugihara, Y., Eljamal, O., 2022. Chemical deposition of iron nanoparticles (FeO) on titanium nanowires for efficient adsorption of ciprofloxacin from water. *Water Practice and Technology* 17(1), 75-83. https://doi.org/10.2166/wpt.2021.091.
- Falyouna, O., Maamoun, I., Bensaida, K., Tahara, A., Sugihara, Y., Eljamal, O., 2022. Encapsulation of iron nanoparticles with magnesium hydroxide shell for remarkable removal of ciprofloxacin from contaminated water. *Journal of Colloid and Interface Science* 605, 813-827. https://doi.org/10.1016/j.jcis.2021.07.154.
- Fan, M., Zhou, N., Li, P., Chen, L., Chen, Y., Shen, S., Zhu, S., 2017. Anaerobic co-metabolic biodegradation of tetrabromobisphenol A using a bioelectrochemical system. *Journal*

of Hazardous Materials 321, 791-800. https://doi.org/10.1016/j.jhazmat.2016.09.068.

- Fan, P., Sun, Y., Qiao, J., Lo, I.M.C., Guan, X., 2018. Influence of weak magnetic field and tartrate on the oxidation and sequestration of Sb(III) by zerovalent iron: Batch and semi-continuous flow study. *Journal of Hazardous Materials* 343, 266-275. https://doi.org/10.1016/j.jhazmat.2017.09.041.
- Fan, Q., Gong, T., Dong, Q., Wang, W., 2023. Uncovering hydrothermal treatment of per- and polyfluoroalkyl substances. *Eco-Environment & Health* 2(1), 21-23. https://doi.org/10.1016/j.eehl.2023.02.002.
- Fan, Y., Liu, W., Sun, Z., Chowwanonthapunya, T., Zhao, Y., Dong, B., Zhang, T., Banthukul, W., 2021. Effect of chloride ion on corrosion resistance of Ni-advanced weathering steel in simulated tropical marine atmosphere. *Construction and Building Materials* 266, 120937. https://doi.org/10.1016/j.conbuildmat.2020.120937.
- Feiteiro, J., Mariana, M., Cairrao, E., 2021. Health toxicity effects of brominated flame retardants: from environmental to human exposure. *Environmental Pollution* 285, 117475. https://doi.org/10.1016/j.envpol.2021.117475.
- Fendorf, S., Michael, H. A., Van Geen, A., 2010. Spatial and temporal variations of groundwater arsenic in South and Southeast Asia. *Science*, 1123-1127.
- Feng, M., Li, H., You, S., Zhang, J., Lin, H., Wang, M., Zhou, J., 2019. Effect of hexavalent chromium on the biodegradation of tetrabromobisphenol A (TBBPA) by Pycnoporus sanguineus. *Chemosphere* 235, 995-1006. https://doi.org/10.1016/j.chemosphere.2019.07.025.
- Feng, P., Guan, X., Sun, Y., Choi, W., Qin, H., Wang, J., Qiao, J., Li, L., 2015. Weak magnetic field accelerates chromate removal by zero-valent iron. *Journal of Environmental Sciences* 31, 175-183. https://doi.org/10.1016/j.jes.2014.10.017.
- Fernández-Martínez, A., Charlet, L., 2009. Selenium environmental cycling and bioavailability: a structural chemist point of view. *Reviews in Environmental Science and Bio/Technology* 8(1), 81-110. https://doi.org/10.1007/s11157-009-9145-3.
- Ferronato, C., Silva, B., Costa, F., Tavares, T., 2016. Vermiculite bio-barriers for Cu and Zn remediation: an eco-friendly approach for freshwater and sediments protection. *International Journal of Environmental Science and Technology* 13(5), 1219-1228. https://doi.org/10.1007/s13762-016-0957-8.
- Filella, M., Belzile, N., Chen, Y.W., 2002. Antimony in the environment a review focused on natural waters I. Occurrence. *Earth Science Reviews* 57(1), 125-176.
- Filella, M., Williams, P.A., 2012. Antimony interactions with heterogeneous complexants in waters, sediments and soils: a review of binding data for homologous compounds. *Geochemistry* 72, 49-65. https://doi.org/10.1016/j.chemer.2012.01.006.
- Fricker, A.D., LaRoe, S.L., Shea, M.E., Bedard, D.L., 2014. Dehalococcoides mccartyi strain JNA dechlorinates multiple chlorinated phenols including pentachlorophenol and harbors

at least 19 reductive dehalogenase homologous genes. *Environmental Science & Technology* 48(24), 14300–14308. https://doi.org/10.1021/es503553f.

- Galdames, A., Ruiz-Rubio, L., Orueta, M., Sanchez-Arzalluz, M., Vilas-Vilela, J.L., 2020. Zero-valent iron nanoparticles for soil and groundwater remediation. *International Journal of Environmental Research and Public Health* 17(16), 5817–5839. https://doi.org/10.3390/ijerph17165817.
- Galloway, J.E., Moreno, A.V.P., Lindstrom, A.B., Strynar, M.J., Newton, S., May, A. A., Weavers, L.K., 2020. Evidence of air dispersion: HFPO-DA and PFOA in Ohio and West Virginia surface water and soil near a fluoropolymer production facility. *Environmental Science & Technology* 54(12), 7175–7184. https://doi.org/10.1021/acs.est.9b07384.
- Genuchten, M.T.V., 1981. Non-equilibrium transport parameters from miscible displacement experiments. *United States Department of Agriculture Science And Education Administration U.S. Salinity Laboratory Riverside*, California, p. 119.
- Gerlach, R., Cunningham, A.B., Caccavo, F., 2000. Dissimilatory iron-reducing bacteria can influence the reduction of carbon tetrachloride by iron metal. *Environmental Science & Technology* 34(12), 2461–2464. https://doi.org/10.1021/es991200h.
- Gholami, F., Mosmeri, H., Shavandi, M., Dastgheib, S.M.M., Amoozegar, M.A., 2019. Application of encapsulated magnesium peroxide (MgO2) nanoparticles in permeable reactive barrier (PRB) for naphthalene and toluene bioremediation from groundwater. *Science of the Total Environment* 655, 633–640. https://doi.org/10.1016/j.scitotenv.2018.11.253.
- Gillespie, I. M. M., Philp, J. C., 2013. Bioremediation, an environmental remediation technology for the bioeconomy. *Trends in Biotechnology* 31(6), 329–332. https://doi.org/https://doi.org/10.1016/j.tibtech.2013.01.015.
- Gogoi, A., Mazumder, P., Tyagi, V.K., Tushara Chaminda, G.G., An, A.K., Kumar, M., 2018. Occurrence and fate of emerging contaminants in water environment: a review. *Groundwater for Sustainable Development* 6, 169–180. https://doi.org/10.1016/j.gsd.2017.12.009.
- Goldman, P., 1965. The enzymatic cleavage of the carbon-fluorine bond in fluoroacetate. *Journal of Biological Chemistry* 240(8), 3434–3438. https://doi.org/10.1016/s0021-9258(18)97236-4.
- Gong, R., Ai, C., Zhang, B., Cheng, X., 2018. Effect of selenite on organic selenium speciation and selenium bioaccessibility in rice grains of two Se-enriched rice cultivars. *Food Chemistry* 264, 443–448. https://doi.org/10.1016/j.foodchem.2018.05.066.
- Gore, F., Fawell, J., Bartram, J., 2010. Too much or too little? A review of the conundrum of selenium. *Journal of Water and Health* 8(3), 405–416. https://doi.org/10.2166/wh.2009.060.
- Gu, B., Phelps, T.J., Liang, L., Dickey, M.J., Roh, Y., Kinsall, B.L., Palumbo, A.V.,

Jacobs, G.K., 1999. Biogeochemical dynamics in zero-valent iron columns: implications for permeable reactive barriers. *Environmental Science & Technology* 33(13), 2170–2177. https://doi.org/10.1021/es981077e.

- Gu, C., Wang, J., Guo, M., Sui, M., Lu, H., Liu, G., 2018. Extracellular degradation of tetrabromobisphenol A via biogenic reactive oxygen species by a marine Pseudoalteromonas sp. *Water Research* 142, 354–362. https://doi.org/10.1016/j.watres.2018.06.012.
- Gu, J., Chen, X., Wang, Y., Wang, L., Szlavecz, K., Ma, Y., Ji, R., 2020. Bioaccumulation, physiological distribution, and biotransformation of tetrabromobisphenol a (TBBPA) in the geophagous earthworm Metaphire guillelmihint for detoxification strategy. *Journal of Hazardous Materials* 388, 122027. https://doi. org/10. 1016/j. jhazmat. 2020. 122027.
- Guan, X., Sun, Y., Qin, H., Li, J., Lo, I.M., He, D., Dong, H., 2015. The limitations of applying zero-valent iron technology in contaminants sequestration and the corresponding countermeasures: the development in zero-valent iron technology in the last two decades (1994–2014). *Water Research* 75, 224–248. https://doi.org/10.1016/j.watres.2015.02.034.
- Guan, X.H., Sun, Y.K., Qin, H.J., Li, J.X., Lo, I.M.C., He, D., Dong, H.R., 2015. The limitations of applying zero-valent iron technology in contaminants sequestration and the corresponding countermeasures: the development in zero-valent iron technology in the last two decades (1994–2014). *Water Research* 75, 224–248. https://doi.org/10.1016/j.watres.2015. 02.034.
- Guo, W., Zhao, Q., Du, J., Wang, H., Li, X., Ren, N., 2020. Enhanced removal of sulfadiazine by sulfidated ZVI activated persulfate process: Performance, mechanisms and degradation pathways. *Chemical Engineering Journal* 388, 124303. https://doi.org/10.1016/j.cej.2020.124303.
- Guo, X., Wu, Z., He, M., 2009. Removal of antimony(V) and antimony(III) from drinking water by coagulation-flocculation-sedimentation (CFS). *Water Research* 43(17), 4327–4335. https://doi.org/10.1016/j.watres.2009.06.033.
- Han, W., Gao, Y., Yao, Q., Yuan, T., Wang, Y., Zhao, S., Shi, R., Bonefeld-Jorgensen, E.C., Shen, X., Tian, Y., 2018. Perfluoroalkyl and polyfluoroalkyl substances in matched parental and cord serum in Shandong, China. *Environment International* 116, 206–213. https://doi.org/10.1016/j.envint.2018.04.025.
- Harrad, S., Wemken, N., Drage, D. S., Abdallah, M. A., Coggins, A. M., 2019. Perfluoroalkyl substances in drinking water, indoor air and dust from ireland: Implications for human exposure. *Environmental Science & Technology* 53(22), 13449–13457. https://doi.org/10.1021/acs.est.9b04604.
- Hartmann, J., van der Aa, M., Wuijts, S., de Roda Husman, A.M., van der Hoek, J.P., 2018. Risk governance of potential emerging risks to drinking water quality: analysing current

practices. *Environmental Science & Policy* 84, 97–104. https://doi.org/10.1016/j.envsci.2018.02.015.

- Hawrylak-Nowak, B., Matraszek, R., Pogorzelec, M., 2015. The dual effects of two inorganic selenium forms on the growth, selected physiological parameters and macronutrients accumulation in cucumber plants. *Acta Physiologiae Plantarum* 37(2), 41–54. https://doi.org/10.1007/s11738-015-1788-9.
- Hayashi, H., Kanie, K., Shinoda, K., Muramatsu, A., Suzuki, S., Sasaki, H., 2009. pH-dependence of selenate removal from liquid phase by reductive Fe(II)-Fe(III) hydroxysulfate compound, green rust. *Chemosphere* 76(5), 638–643. https://doi.org/10.1016/j.chemosphere.2009.04.037.
- He, F., Zhao, D., Paul, C., 2010. Field assessment of carboxymethyl cellulose stabilized iron nanoparticles for in situ destruction of chlorinated solvents in source zones. *Water Research* 44(7), 2360–2370. https://doi.org/10.1016/j.watres.2009.12.041.
- He, M., 2007. Distribution and phytoavailability of antimony at an antimony mining and smelting area, Hunan, China. *Environmental Geochemistry and Health* 29(3), 209–219. https://doi.org/10.1007/s10653-006-9066-9.
- He, M., Wang, X., Wu, F., Fu, Z., 2012. Antimony pollution in China. *Science of the Total Environment* 421–422, 41–50. https://doi.org/10.1016/j.scitotenv.2011.06.009.
- He, Y., Xiang, Y., Zhou, Y., Yang, Y., Zhang, J., Huang, H., Shang, C., Luo, L., Gao, J., Tang, L., 2018. Selenium contamination, consequences and remediation techniques in water and soils: a review. *Environmental Research* 164, 288–301. https://doi.org/10.1016/j.envres.2018.02.037.
- Hellsing., M.S., Josefsson., S., Hughes., A.V., ., L.A., 2016. Sorption of perfluoroalkyl substances to two types of minerals. *Chemosphere* 159, 385–391. https://doi.org/10.1016/j.chemosphere.2016.06.016.
- Holliger, C.S., G.; Stams,A.J.M.; Zehnder,A.J.B. , 1990. Reductive dechlorination of 1,2-dichloroethane and chloroethane by cell suspensions of methanogenic bacteria. *Biodegradation* 1, 253–261.
- Hu, X., He, M., 2017. Organic ligand-induced dissolution kinetics of antimony trioxide. *Journal of Environmental Sciences* 56, 87–94. https://doi.org/10.1016/j.jes.2016.09.006.
- Huang, B.B., Lei, C., Wei, C.H., Zeng, G.M., 2014. Chlorinated volatile organic compounds (Cl-VOCs) in environment-sources, potential human health impacts, and current remediation technologies. *Environment International* 71, 118–138. https://doi.org/10.1016/j.envint.2014.06.013.
- Huang, D., Xu, R., Sun, X., Li, Y., Xiao, E., Xu, Z., Wang, Q., Gao, P., Yang, Z., Lin, H., Sun, W., 2022. Effects of perfluorooctanoic acid (PFOA) on activated sludge microbial community under aerobic and anaerobic conditions. *Environmental Science and*

Pollution Research 29, 63379–63392. https://doi.org/10.1007/s11356-022-18841-8.

- Huang, J.H., 2018. Characterising microbial reduction of arsenate sorbed to ferrihydrite and its concurrence with iron reduction. *Chemosphere* 194, 49–56. https://doi.org/10.1016/j.chemosphere.2017.11.109.
- Huang, L., Liu, G., Dong, G., Wu, X., Wang, C., Liu, Y., 2017. Reaction mechanism of zero-valent iron coupling with microbe to degrade tetracycline in permeable reactive barrier (PRB). *Chemical Engineering Journal* 316, 525–533. https://doi.org/10.1016/j.cej.2017.01.096.
- Huang, P., Ye, Z., Xie, W., Chen, Q., Li, J., Xu, Z., Yao, M., 2013. Rapid magnetic removal of aqueous heavy metals and their relevant mechanisms using nanoscale zero valent iron (nZVI) particles. *Water Research* 47(12), 4050–4058. https://doi.org/10.1016/j.watres.2013.01.054.
- Huang, T., Zhang, G., Zhang, N., Ye, J., Xian, G., 2018. Pre-magnetization by weak magnetic field enhancing Fe^0-Fenton process for wastewater treatment. *Chemical Engineering Journal* 346, 120–126. https://doi.org/10.1016/j.cej.2018.04.009.
- Huang, W., Yang, F., Huang, W., Wang, D., Lei, Z., Zhang, Z., 2019. Weak magnetic field significantly enhances methane production from a digester supplemented with zero valent iron. *Bioresource Technology* 282, 202–210. https://doi.org/10.1016/j.biortech.2019.03.013.
- Huang, Y., Zeng, Q., Hu, L., Xiong, D., Zhong, H., He, Z., 2021. Column study of enhanced Cr(VI) removal and removal mechanisms by Sporosarcina saromensis W5 assisted bio-permeable reactive barrier. *Journal of Hazardous Materials* 405, 124115. https://doi.org/10.1016/j.jhazmat.2020.124115.
- Huang, Y., Zeng, Q., Hu, L., Zhong, H., He, Z., 2022. Column study of enhanced Cr(VI) removal by bio-permeable reactive barrier constructed from novel iron-based material and Sporosarcina saromensis W5. *Environmental Science and Pollution Research* 29, 44893–44905. https://doi.org/10.1007/s11356-022-18972-y.
- Inglezakis, V. J., Malamis, S., Omirkhan, A., Nauruzbayeva, J., Makhtayeva, Z., Seidakhmetov, T., Kudarova, A., 2017. Investigating the inhibitory effect of cyanide, phenol and 4-nitrophenol on the activated sludge process employed for the treatment of petroleum wastewater. *Journal of Environmental Management* 203(Pt 2), 825–830. https://doi.org/10.1016/j.jenvman.2016.08.066.
- Jarvinen, K. T., Melin, E. S., Puhakka, J. A., 1994. High-rate bioremediation of chlorophenol-contaminated groundwater at low temperatures. *Environmental Science & Technology* 28(13), 2387–2392. https://doi.org/10.1021/es00062a025.
- Jeon, J., Kim, C., Kim, H., Lee, C., Hwang, S., Choi, S., 2021. Spatial distribution, source identification, and anthropogenic effects of brominated flame retardants in nationwide soil collected from South Korea. *Environmental Pollution* 272, 116026. https://doi.org/10.

1016/j.envpol.2020.116026.

- Ji, Y., Yan, N., Brusseau, M.L., Guo, B., Zheng, X., Dai, M., Liu, H., Li, X., 2021. Impact of a hydrocarbon surfactant on the retention and transport of perfluorooctanoic acid in saturated and unsaturated porous media. *Environmental Science & Technology* 55(15), 10480–10490. https://doi.org/10.1021/acs.est.1c01919.
- Jia, G., Tang, X., Xu, J., 2021. Synthesis of hydrochar supported zero-valent iron composites through hydrothermal carbonization of granatum and zero-valent iron: potential applications for Pb(2+) removal. *Water Science & Technology* 84(8), 1873–1884. https://doi.org/10.2166/wst.2021.366.
- Jiang, J.H., Li, Y.H., Cai, W.M., 2008. Experimental and mechanism research of SO_2 removal by cast iron scraps in a magnetically fixed bed. *Journal of Hazardous Materials* 153 (1–2), 508–513. https://doi.org/10.1016/j.jhazmat.2007.08.083.
- Jiang, X., Qiao, J., Lo, I.M., Wang, L., Guan, X., Lu, Z., Zhou, G., Xu, C., 2015. Enhanced paramagnetic Cu(2)(+) ions removal by coupling a weak magnetic field with zero valent iron. *Journal of Hazardous Materials* 283, 880–887. https://doi.org/10.1016/j.jhazmat.2014.10.044.
- Johnson, G.R., 2022. PFAS in soil and groundwater following historical land application of biosolids. *Water Research* 211, 118035. https://doi.org/10.1016/j.watres.2021.118035.
- Joudan., S., Lundgren., R.J., 2022. Taking the "F" out of forever chemicals. *Science* 377 (6608), 816–817.
- Ju, K.S., Parales, R.E., 2010. Nitroaromatic compounds, from synthesis to biodegradation. *Microbiology and Molecular Biology Reviews: MMBR* 74(2), 250–272. https://doi.org/10.1128/MMBR.00006-10.
- Kalhor, K., Ghasemizadeh, R., Rajic, L., Alshawabkeh, A., 2019. Assessment of groundwater quality and remediation in karst aquifers: a review. *Groundwater for Sustainable Development* 8, 104–121. https://doi.org/10.1016/j.gsd.2018.10.004.
- Kamra, S.K., Lennartz, B., Genuchten, M.T.V., Widmoser, P., 2001. Evaluating non-equilibrium solute transport in small soil column. *Journal of Contaminant Hydrology* 48, 189–212.
- Kamran, U., Bhatti, H.N., Iqbal, M., Jamil, S., Zahid, M., 2019. Biogenic synthesis, characterization and investigation of photocatalytic and antimicrobial activity of manganese nanoparticles synthesized from Cinnamomum verum bark extract. *Journal of Molecular Structure* 1179, 532–539. https://doi.org/10.1016/j.molstruc.2018.11.006.
- Kamran, U., Heo, Y.-J., Min, B.-G., In, I., Park, S.-J., 2020. Effect of nickel ion doping in MnO_2/reduced graphene oxide nanocomposites for lithium adsorption and recovery from aqueous media. *RSC Advances* 10(16), 9245–9257. https://doi.org/10.1039/c9ra10277a.
- Kamran, U., Park, S.-J., 2020. Microwave-assisted acid functionalized carbon nanofibers

decorated with Mn doped TNTs nanocomposites: Efficient contenders for lithium adsorption and recovery from aqueous media. *Journal of Industrial and Engineering Chemistry* 92, 263-277. https://doi.org/10.1016/j.jiec.2020.09.014.

- Kang, M., Kawasaki, M., Tamada, S., Kamei, T., Magara, Y., 2001. Effect of pH on the removal of arsenic and antimony using reverse osmosis membranes. *Desalination* 131, 293-298.
- Kang, N., Zhu, N., Guo, W., Shi, C., Wu, P., Wei, X., 2018. Efficient debromination of Tetrabromobisphenol A (TBBPA) by Au/Fe @ biocarbon derived from bioreduction precious metals. *Chemical Engineering Journal* 334, 99-107. https://doi.org/10.1016/j.cej.2017.10.018.
- Karimipour, Z., Jalilzadeh Yengejeh, R., Haghighatzadeh, A., Mohammadi, M. K., Mohammadi Rouzbehani, M., 2021. UV-induced photodegradation of 2,4,6-trichlorophenol using Ag-Fe_2O_3-CeO_2 photocatalysts. *Journal of Inorganic and Organometallic Polymers and Materials* 31(3), 1143-1152. https://doi.org/10.1007/s10904-020-01859-1.
- Kieliszek, M., Blazejak, S., 2016. Current knowledge on the importance of Selenium in food for living organisms: a review. *Molecules* 21(5), 609-625. https://doi.org/10.3390/molecules21050609.
- Kim, D.-h., Kim, J., Choi, W., 2011. Effect of magnetic field on the zero valent iron induced oxidation reaction. *Journal of Hazardous Materials* 192(2), 928-931. https://doi.org/https://doi.org/10.1016/j.jhazmat.2011.05.075.
- Kim, E. J., Kim, J. H., Azad, A. M., Chang, Y. S., 2011. Facile synthesis and characterization of Fe/FeS nanoparticles for environmental applications. *ACS Appl Mater Interfaces* 3(5), 1457-1462. https://doi.org/10.1021/am200016v.
- Kim, M. H., Wang, N., McDonald, T., Chu, K. H., 2012. Biodefluorination and biotransformation of fluorotelomer alcohols by two alkane-degrading Pseudomonas strains. *Biotechnology and Bioengineering* 109(12), 3041-3048. https://doi.org/10.1002/bit.24561.
- Kim, N., Park, M., Yun, Y.S., Park, D., 2019. Removal of anionic arsenate by a PEI-coated bacterial biosorbent prepared from fermentation biowaste. *Chemosphere* 226, 67-74. https://doi.org/10.1016/j.chemosphere.2019.03.113.
- Kim, N.K., Lee, S.H., Yoon, H., Jeong, G., Jung, Y.J., Hur, M., Lee, B.H., Park, H. D., 2021. Microbiome degrading linear alkylbenzene sulfonate in activated sludge. *Journal of Hazardous Materials* 418, 126365. https://doi.org/10.1016/j.jhazmat.2021.126365.
- Kim, Y., Carraway, E., 2000. Dechlorination of pentachlorophenol by zero-valent iron and modified zero-valent irons. *Environmental Science & Technology* 34, 2014-2017.
- Kornilovych, B., Wireman, M., Ubaldini, S., Guglietta, D., Koshik, Y., Caruso, B., Kovalchuk, I., 2018. Uranium removal from groundwater by permeable reactive barrier with zero-valent iron and organic carbon mixtures: laboratory and field studies. *Metals* 8(6),

408–423. https://doi.org/10.3390/met8060408.

- Kumar, N., Couture, R.M., Millot, R., Battaglia-Brunet, F., Rose, J., 2016. Microbial sulfate reduction enhances arsenic mobility downstream of zerovalent-iron-based permeable reactive barrier. *Environmental Science & Technology* 50(14), 7610–7617. https://doi.org/10.1021/acs.est.6b00128.
- Kumari, A., Maurya, N.S., 2019. Arsenic contamination and associated health risk (brief review). *Oriental Journal of Chemistry* 35(2), 563–570. https://doi.org/10.13005/ojc/350209.
- Kumari, M., Ghosh, P., Thakur, I.S., 2018. Application of microbes in remediation of hazardous wastes: a review, in: Varjani, S. J., Agarwal, A. K., Gnansounou, E., Gurunathan, B. (Eds.), *Bioremediation: Applications for Environmental Protection and Management*. Springer Singapore, Singapore, 223–241. https://doi.org/10.1007/978-981-10-7485-1_11.
- Kursvietiene, L., Mongirdiene, A., Bernatoniene, J., Sulinskiene, J., Staneviciene, I., 2020. Selenium anticancer properties andimpact on cellular redox status. *Antioxidants* 9(1), 80–91. https://doi.org/10.3390/antiox9010080.
- Lai, B., Zhang, Y., Chen, Z., Yang, P., Zhou, Y., Wang, J., 2014. Removal of p-nitrophenol (PNP) in aqueous solution by the micron-scale iron-copper (Fe/Cu) bimetallic particles. *Applied Catalysis B: Environmental* 144, 816–830. https://doi.org/10.1016/j.apcatb.2013.08.020.
- Lai, K.C.K., Lo, I.M.C., 2008. Removal of chromium (VI) by acid-washed zero-valent iron under various groundwater geochemistry conditions. *Environmental Science & Technology* 42(4), 1238–1244. https://doi.org/10.1021/es071572n.
- Lapworth, D.J., Lopez, B., Laabs, V., Kozel, R., Wolter, R., Ward, R., Vargas Amelin, E., Besien, T., Claessens, J., Delloye, F., Ferretti, E., Grath, J., 2019. Developing a groundwater watch list for substances of emerging concern: a European perspective. *Environmental Research Letters* 14(3), 35004–35018. https://doi.org/10.1088/1748-9326/aaf4d7.
- Law, R.J., Allchin, C.R., Boer, J.d., Covaci, A., Herzke, D., Lepom, P., Morris, S., Tronczynski, J., Wit, C.A.d., 2006. Levels and trends of brominated flame retardants in the European environment. *Chemosphere* 64, 187–208. https://doi.org/10.1016/j.chemosphere.2005.12.007.
- Lawal Wasiu, A., Choi, H., 2018. Feasibility study on the removal of perfluorooctanoic acid by using palladium-doped nanoscale zerovalent iron. *Journal of Environmental Engineering* 144(11), 04018115. https://doi.org/10.1061/(ASCE)EE.1943-7870.0001468.
- Lawrinenko, M., Kurwadkar, S., Wilkin, R.T., 2023. Long-term performance evaluation of zero-valent iron amended permeable reactive barriers for groundwater remediation — a mechanistic approach. *Geosci Front* 14(2), 1–13. https://doi.org/10.1016/j.gsf.2022.

101494.

- Lee, Y.C., Chen, M.J., Huang, C.P., Kuo, J., Lo, S.L., 2016. Efficient sonochemical degradation of perfluorooctanoic acid using periodate. *Ultrasonics Sonochemistry* 31, 499–505. https://doi.org/10.1016/j.ultsonch.2016.01.030.
- Lemic, D., Orešković, M., Mikac, K.M., Marijan, M., Jurić, S., Vlahoviček-Kahlina, K., Vinceković, M., 2021. Sustainable pest management using biodegradable apitoxin-loaded calcium-alginate microspheres. *Sustainability* 13 (11), 6167. https://doi. org/10. 3390/su13116167.
- Leng, Y.Q., Guo, W.L., Su, S.N., Yi, C.L., Xing, L.T., 2012. Removal of antimony (III) from aqueous solution by graphene as an adsorbent. *Chemical Engineering Journal* 211, 406–411.
- Li, C., Yang, L., Shi, M., Liu, G., 2019. Persistent organic pollutants in typical lake ecosystems. *Ecotoxicology and Environmental Safety* 180, 668–678. https://doi.org/10.1016/j.ecoenv.2019.05.060.
- Li, D., Mao, Z., Zhong, Y., Huang, W., Wu, Y., Peng, P. a., 2016. Reductive transformation of tetrabromobisphenol A by sulfidated nano zerovalent iron. *Water Research* 103, 1–9. https://doi.org/10.1016/j.watres.2016.07.003.
- Li, F., Jiang, B., Nastold, P., Kolvenbach, B. A., Chen, J., Wang, L., Guo, H., Corvini, P.F., Ji, R., 2015. Enhanced transformation of tetrabromobisphenol a by nitrifiers in nitrifying activated sludge. *Environmental Science&Technology* 49(7), 4283–4292. https://doi.org/10.1021/es5059007.
- Li, H., Cao, W., Wang, W., Huang, Y., Xiang, M., Wang, C., Chen, S., Si, R., Huang, M., 2021. Carbon nanotubes mediating nano alpha-FeOOH reduction by Shewanella putrefaciens CN32 to enhance tetrabromobisphenol A removal. *Science of the Total Environment* 777, 146183. https://doi.org/10.1016/j.scitotenv.2021.146183.
- Li, H., Chen, S., Ren, L.Y., Zhou, L.Y., Tan, X.J., Zhu, Y., Belver, C., Bedia, J., Yang, J., 2019 Biochar mediates activation of aged nanoscale ZVI by Shewanella putrefaciens CN32 to enhance the degradation of Pentachlorophenol. *Chemical Engineering Journal* 368, 148–156. https://doi.org/10.1016/j.cej.2019.02.099.
- Li, H., Liu, Q., 2022. Reaction medium for permeable reactive barrier remediation of groundwater polluted by heavy metals. *Frontiers in Environmental Science* 10, 968546. https://doi.org/10.3389/fenvs.2022.968546.
- Li, J., Bao, H., Xiong, X., Sun, Y., Guan, X., 2015. Effective Sb(V) immobilization from water by zero-valent iron with weak magnetic field. *Separation and Purification Technology* 151, 276–283. https://doi.org/10.1016/j.seppur.2015.07.056.
- Li, J., Shi, Z., Ma, B., Zhang, P., Jiang, X., Xiao, Z., Guan, X., 2015. Improving the reactivity of zerovalent iron by taking advantage of its magnetic memory: Implications for

arsenite removal. *Environmental Science & Technology* 49(17), 10581–10588. https://doi.org/10.1021/acs.est.5b02699.

- Li, J., Zhang, B., 2020. Woodchip-sulfur packed biological permeable reactive barrier for mixotrophic vanadium (V) detoxification in groundwater. *Science China Technological Sciences* 63(11), 2283–2291. https://doi.org/10.1007/s11431-020-1655-6.
- Li, X., Zhou, M., Pan, Y., 2018. Enhanced degradation of 2,4-dichlorophenoxyacetic acid by pre-magnetization Fe-C activated persulfate: influential factors, mechanism and degradation pathway. *Journal of Hazardous Materials* 353, 454–465. https://doi.org/10.1016/j.jhazmat.2018.04.035.
- Li, X., Zhou, M., Pan, Y., Xu, L., 2017a. Pre-magnetized Fe^0/persulfate for notably enhanced degradation and dechlorination of 2, 4-dichlorophenol. *Chemical Engineering Journal* 307, 1092–1104. https://doi.org/10.1016/j.cej.2016.08.140.
- Li, X., Zhou, M., Pan, Y., Xu, L., Tang, Z., 2017b. Highly efficient advanced oxidation processes (AOPs) based on pre-magnetization Fe^0 for wastewater treatment. *Separation and Purification Technology* 178, 49–55. https://doi.org/10.1016/j.seppur.2016.12.050.
- Li, Y., Huang, Y., Wu, W., Yan, M., Xie, Y., 2021. Research and application of arsenic-contaminated groundwater remediation by manganese ore permeable reactive barrier. *Environmental Technology* 42(13), 2009–2020. https://doi.org/10.1080/09593330.2019.1687587.
- Li, Y., Oliver, D.P., Kookana, R.S., 2018. A critical analysis of published data to discern the role of soil and sediment properties in determining sorption of per and polyfluoroalkyl substances (PFASs). *Science of the Total Environment* 628–629, 110–120. https://doi.org/10.1016/j.scitotenv.2018.01.167.
- Li, Z., Brusseau, M.L., 2000. Nonideal transport of reactive solutes in heterogeneous porous media: 6. Microscopic and macroscopic approaches for incorporating heterogeneous rate-limited mass transfer. *Water Resources Research* 36(10), 2853–2867.
- Li, Z., Lyu, X., Gao, B., Xu, H., Wu, J., Sun, Y., 2021. Effects of ionic strength and cation type on the transport of per fluorooctanoic acid (PFOA) in unsaturated sand porous media. *Journal of Hazardous Materials* 403, 123688. https://doi.org/10.1016/j.jhazmat.2020.123688.
- Li, Z., Xu, S., Xiao, G., Qian, L., Song, Y., 2019. Removal of hexavalent chromium from groundwater using sodium alginate dispersed nano zero-valent iron. *Journal of Environmental Management* 244, 33–39. https://doi.org/https://doi.org/10.1016/j.jenvman.2019.04.130.
- Liang, D., Liu, X., Woodard, T.L., Holmes, D.E., Smith, J.A., Nevin, K.P., Feng, Y., Lovley, D.R., 2021. Extracellular electron exchange capabilities of desulfovibrio ferrophilus and desulfopila corrodens. *Environmental Science & Technology* 55, 16195–16203.

https://doi.org/10.1021/acs.est.1c04071.

- Liang, L., Guan, X., Huang, Y., Ma, J., Sun, X., Qiao, J., Zhou, G., 2015. Efficient selenate removal by zero-valent iron in the presence of weak magnetic field. *Separation and Purification Technology* 156, 1064-1072. https://doi.org/10.1016/j.seppur.2015.09.062.
- Liang, L., Guan, X., Shi, Z., Li, J., Wu, Y., Tratnyek, P.G., 2014a. Coupled effects of aging and weak magnetic fields on sequestration of selenite by zero-valent iron. *Environmental Science & Technology* 48(11), 6326-6334. https://doi.org/10.1021/es500958b.
- Liang, L., Sun, W., Guan, X., Huang, Y., Choi, W., Bao, H., Li, L., Jiang, Z., 2014b. Weak magnetic field significantly enhances selenite removal kinetics by zero valent iron. *Water Research* 49, 371-380. https://doi.org/10.1016/j.watres.2013.10.026.
- Lim, X., 2019. Tainted water: the scientists tracing thousands of fluorinated chemicals in our environment. *Nature*, 26-29.
- Lin, X.Q., Li, Z.L., Liang, B., Zhai, H.L., Cai, W.W., Nan, J., Wang, A.J., 2019. Accelerated microbial reductive dechlorination of 2,4,6-trichlorophenol by weak electrical stimulation. *Water Research* 162, 236-245. https://doi.org/10.1016/j.watres.2019.06.068.
- Lin, X.Q., Li, Z.L., Zhu, Y.Y., Chen, F., Liang, B., Nan, J., Wang, A.J., 2020. Palladium/iron nanoparticles stimulate tetrabromobisphenol a microbial reductive debromination and further mineralization in sediment. *Environment International* 135, 105353. https://doi.org/10.1016/j.envint.2019.105353.
- Litter, M.I., Morgada, M.E., Bundschuh, J., 2010. Possible treatments for arsenic removal in Latin American waters for human consumption. *Environmental Pollution* 158(5), 1105-1118. https://doi.org/10.1016/j.envpol.2010.01.028.
- Liu, C., Chen, X., Banwart, S.A., Du, W., Yin, Y., Guo, H., 2021. A novel permeable reactive biobarrier for ortho-nitrochlorobenzene pollution control in groundwater: Experimental evaluation and kinetic modelling. *Journal of Hazardous Materials* 420, 126563. https://doi.org/10.1016/j.jhazmat.2021.126563.
- Liu, C., Chen, X., Mack, E.E., Wang, S., Du, W., Yin, Y., Banwart, S.A., Guo, H., 2019. Evaluating a novel permeable reactive bio-barrier to remediate PAH-contaminated groundwater. *Journal of Hazardous Materials* 368, 444-451. https://doi.org/10.1016/j.jhazmat.2019.01.069.
- Liu, G., Feng, M., Tayyab, M., Gong, J., Zhang, M., Yang, M., Lin, K., 2021. Direct and efficient reduction of perfluorooctanoic acid using bimetallic catalyst supported on carbon. *Journal of Hazardous Materials* 412, 125224. https://doi.org/10.1016/j.jhazmat.2021.125224.
- Liu, G., Tang, H., Fan, J., Xie, Z., He, T., Shi, R., Liao, B., 2019. Removal of 2,4,6-trichlorophenol from water by Eupatorium adenophorum biochar-loaded nano-iron/nickel. *Bioresoure Technology* 289, 121734. https://doi.org/10.1016/j.biortech.2019.121734.

- Liu, G., Tang, H., Fan, J., Xie, Z., He, T., Shi, R., Liao, B., 2019. Removal of 2,4, 6-trichlorophenol from water by Eupatorium adenophorum biochar-loaded nano-iron/nickel. *Bioresource Technology* 289, 121734. https://doi.org/10.1016/j.biortech.2019.121734.
- Liu, H., Chen, S., Lu, J., Li, Q., Li, J., Zhang, B., 2022. Pentavalent vanadium and hexavalent uranium removal from groundwater by woodchip-sulfur based mixotrophic biotechnology. *Chemical Engineering Journal* 437, 135313. https://doi.org/10.1016/j.cej.2022.135313.
- Liu, J., Qu, R., Wang, Z., Mendoza-Sanchez, I., Sharma, V.K., 2017. Thermal- and photo-induced degradation of perfluorinated carboxylic acids: Kinetics and mechanism. *Water Research* 126, 12–18. https://doi.org/10.1016/j.watres.2017.09.003.
- Liu, K., Li, J., Yan, S., Zhang, W., Li, Y., Han, D., 2016. A review of status of tetrabromobisphenol A (TBBPA) in China. *Chemosphere* 148, 8–20. https://doi.org/10.1016/j.chemosphere.2016.01.023.
- Liu, N., Ding, F., Wang, L., Liu, P., Yu, X., Ye, K., 2016. Coupling of bio-PRB and enclosed in-well aeration system for remediation of nitrobenzene and aniline in groundwater. *Environmental Science and Pollution Research International* 23(10), 9972–9983. https://doi.org/10.1007/s11356-016-6206-3.
- Liu, N., Zhang, Y., An, Y., Wang, L., 2018. Preparation of integrative cubes as a novel biological permeable reactive barrier medium for the enhancement of in situ aerobic bioremediation of nitrobenzene-contaminated groundwater. *Environmental Earth Sciences* 77 (19), 707–716. https://doi.org/10.1007/s12665-018-7890-8.
- Liu, Y., Lowry, G.V., 2006. Effect of particle age (Fe^0 content) and solution pH on NZVI reactivity: H_2 evolution and TCE dechlorination. *Environmental Science & Technology* 40 (19), 6085–6090. https://doi.org/10.1021/es060685o.
- Liu, Y., Ptacek, C.J., Baldwin, R.J., Cooper, J.M., Blowes, D.W., 2020. Application of zero-valent iron coupled with biochar for removal of perfluoroalkyl carboxylic and sulfonic acids from water under ambient environmental conditions. *Science of The Total Environment* 719, 137372. https://doi.org/https://doi.org/10.1016/j.scitotenv.2020.137372.
- Luek, A., Brock, C., Rowan, D.J., Rasmussen, J.B., 2014. A Simplified Anaerobic Bioreactor for the treatment of selenium-laden discharges from non-acidic, end-pit lakes. *Mine Water and the Environment* 33(4), 295–306. https://doi.org/10.1007/s10230-014-0296-2.
- Luo, S., Yang, S., Wang, X., Sun, C., 2010. Reductive degradation of tetrabromobisphenol A over iron-silver bimetallic nanoparticles under ultrasound radiation. *Chemosphere* 79, 672–678. https://doi.org/10.1016/j.chemosphere.2010.02.011.
- Lv, X., Sun, Y., Ji, R., Gao, B., Wu, J., Lu, Q., Jiang, H., 2018. Physicochemical factors controlling the retention and transport of perfluorooctanoic acid (PFOA) in saturated sand and limestone porous media. *Water Research* 141, 251–258. https://doi.org/10.1016/j.

watres.2018.05.020.

- Lv, X., Xue, X., Jiang, G., Wu, D., Sheng, T., Zhou, H., Xu, X., 2014. Nanoscale zero-valent iron (nZVI) assembled on magnetic Fe_3O_4/graphene for chromium (VI) removal from aqueous solution. *Journal of Colloid and Interface Science* 417, 51-59. https://doi.org/10.1016/j.jcis.2013.11.044.
- Lyu, X., Liu, X., Sun, Y., Gao, B., Ji, R., Wu, J., Xue, Y., 2020a. Importance of surface roughness on perfluorooctanoic acid (PFOA) transport in unsaturated porous media. *Environmental Pollution* 266(Pt 1), 115343. https://doi.org/10.1016/j.envpol.2020.115343.
- Lyu, X., Liu, X., Sun, Y., Ji, R., Gao, B., Wu, J., 2019. Transport and retention of perfluorooctanoic acid (PFOA) in natural soils: Importance of soil organic matter and mineral contents, and solution ionic strength. *Journal of Contaminant Hydrology* 225, 103477. https://doi.org/10.1016/j.jconhyd.2019.03.009.
- Lyu, X., Liu, X., Wu, X., Sun, Y., Gao, B., Wu, J., 2020b. Importance of Al/Fe oxyhydroxide coating and ionic strength in perfluorooctanoic acid (PFOA) transport in saturated porous media. *Water Research* 175, 115685. https://doi.org/10.1016/j.watres.2020.115685.
- Lyu, Y., Brusseau, M.L., Chen, W., Yan, N., Fu, X., Lin, X., 2018. Adsorption of PFOA at the Air-Water Interface during Transport in Unsaturated Porous Media. *Environmental Science & Technology* 52(14), 7745-7753. https://doi.org/10.1021/acs.est.8b02348.
- Ma, C., Wu, Y., 2007. Dechlorination of perchloroethylene using zero-valent metal and microbial community. *Environmental Geology* 55(1), 47-54. https://doi.org/10.1007/s00254-007-0963-8.
- Ma, S.-H., Wu, M.-H., Tang, L., Sun, R., Zang, C., Xiang, J.-J., Yang, X.-X., Li, X., Xu, G., 2017. EB degradation of perfluorooctanoic acid and perfluorooctane sulfonate in aqueous solution. *Nuclear Science and Techniques* 28(9), 137-145. https://doi.org/10.1007/s41365-017-0278-8.
- Ma, S., Yue, C., Tang, J., Lin, M., Zhuo, M., Yang, Y., Li, G., An, T., 2021. Occurrence and distribution of typical semi-volatile organic chemicals (SVOCs) in paired indoor and outdoor atmospheric fine particle samples from cities in southern China. *Environmental Pollution* 269, 116123. https://doi.org/10.1016/j.envpol.2020.116123.
- Maamoun, I., Eljamal, O., Falyouna, O., Eljamal, R., Sugihara, Y., 2019. Stimulating effect of magnesium hydroxide on aqueous characteristics of iron nanocomposites. *Water Science & Technology* 80(10), 1996-2002. https://doi.org/10.2166/wst.2020.027.
- Maamoun, I., Eljamal, R., Falyouna, O., Bensaida, K., Sugihara, Y., Eljamal, O., 2021. Insights into kinetics, isotherms and thermodynamics of phosphorus sorption onto nanoscale zero-valent iron. *Journal of Molecular Liquids* 328, 115402. https://doi.org/10.

1016/j.molliq.2021.115402.

- Maĉedo, W., Oliveira, G., Zaiat, M., 2021. Tetrabromobisphenol A (TBBPA) anaerobic biodegradation occurs during acidogenesis. *Chemosphere* 282, 130995. https://doi.org/10.1016/j.chemosphere.2021.130995.
- Madejón, P., Domínguez, M.T., Girón, I., Burgos, P., López-Fernández, M.T., Porras, Ó. G., Madejón, E., 2022. Assessment of the phytoremediation effectiveness in the restoration of uranium mine tailings. *Ecological Engineering* 180, 106669. https://doi.org/10.1016/j.ecoleng.2022.106669.
- Manceau, A., Merkulova, M., Murdzek, M., Batanova, V., Baran, R., Glatzel, P., Saikia, B. K., Paktunc, D., Lefticariu, L., 2018. Chemical forms of mercury in pyrite: Implications for predicting mercury releases in acid mine drainage settings. *Environmental Science & Technology* 52(18), 10286–10296. https://doi.org/10.1021/acs.est.8b02027.
- Marsolek, M. D., Kirisits, M. J., Rittmann, B. E., 2007. Biodegradation of 2, 4, 5-trichlorophenol by aerobic microbial communities: biorecalcitrance, inhibition, and adaptation. *Biodegradation* 18(3), 351–358. https://doi.org/10.1007/s10532-006-9069-3.
- Mayacela-Rojas, C.M., Molinari, A., Cortina, J.L., Gibert, O., Ayora, C., Tavolaro, A., Rivera-Velasquez, M. F., Fallico, C., 2021. Removal of transition metals from contaminated aquifers by PRB technology: Performance comparison among reactive materials. *International Journal of Environmental Research and Public Health* 18(11), 6075–6101. https://doi.org/10.3390/ijerph18116075.
- Mayacela Rojas, C.M., Rivera Velasquez, M.F., Tavolaro, A., Molinari, A., Fallico, C., 2017. Use of vegetable fibers for PRB to remove heavy metals from contaminated aquifers-comparisons among cabuya fibers, broom fibers and ZVI. *International Journal of Environmental Research and Public Health* 14(7), 684–702. https://doi.org/10.3390/ijerph14070684.
- Mazzeo, D. E., Levy, C. E., de Angelis Dde, F., Marin-Morales, M. A., 2010. BTEX biodegradation by bacteria from effluents of petroleum refinery. *Science of Total Environment* 408(20), 4334–4340. https://doi.org/10.1016/j.scitotenv.2010.07.004.
- Mdlovu, N.V., Lin, K.S., Chen, C.Y., Mavuso, F.A., Kunene, S.C., Carrera Espinoza, M. J., 2019. In-situ reductive degradation of chlorinated DNAPLs in contaminated groundwater using polyethyleneimine-modified zero-valent iron nanoparticles. *Chemosphere* 224, 816–826. https://doi.org/10.1016/j.chemosphere.2019.02.160.
- Miao, Q., Li, X., Xu, Y., Liu, C., Xie, R., Lv, Z., 2021. Chemical characteristics of groundwater and source identification in a coastal city. *The Public Library of Science* 16(8), 1–17. https://doi.org/10.1371/journal.pone.0256360.
- Miao, Y., Han, F., Pan, B., Niu, Y., Nie, G., Lv, L., 2014. Antimony(V) removal from water by hydrated ferric oxides supported by calcite sand and polymeric anion exchanger.

Journal of Environmental Sciences 26(2), 307–314. https://doi.org/10.1016/s1001-0742(13)60418-0.

- Minella, M., Bertinetti, S., Hanna, K., Minero, C., Vione, D., 2019. Degradation of ibuprofen and phenol with a Fenton-like process triggered by zero-valent iron (ZVI-Fenton). *Environmental Research* 179(Pt A), 108750. https://doi.org/10.1016/j.envres.2019.108750.
- Morrison, S.J., Metzler, D.R., Dwyer, B.P., 2002. Removal of As, Mn, Mo, Se, U, V and Zn from groundwater by zero-valent iron in a passive treatment cell reaction progress modeling. *Journal of Contaminant Hydrology* 56(1), 99–116.
- Mosmeri, H., Alaie, E., Shavandi, M., Dastgheib, S.M.M., Tasharrofi, S., 2017. Bioremediation of benzene from groundwater by calcium peroxide (CaO_2) nanoparticles encapsulated in sodium alginate. *Journal of the Taiwan Institute of Chemical Engineers* 78, 299–306. https://doi.org/10.1016/j.jtice.2017.06.020.
- Motlagh, A.M., Yang, Z., Saba, H., 2020. Groundwater quality. *Water Environment Research* 92(10), 1649–1658. https://doi.org/10.1002/wer.1412.
- Mukherjee, R., Kumar, R., Sinha, A., Lama, Y., Saha, A.K., 2015. A review on synthesis, characterization, and applications of nano zero valent iron (nZVI) for environmental remediation. *Critical Reviews in Environmental Science and Technology* 46(5), 443–466. https://doi.org/10.1080/10643389.2015.1103832.
- Mwamulima, T., Zhang, X., Wang, Y., Song, S., Peng, C., 2017. Novel approach to control adsorbent aggregation: iron fixed bentonite-fly ash for Lead (Pb) and Cadmium (Cd) removal from aqueous media. *Frontiers of Environmental Science & Engineering* 12(2), 2. https://doi.org/10.1007/s11783-017-0979-6.
- Nam, S.H., Yang, C.Y., An, Y.J., 2009. Effects of antimony on aquatic organisms (Larva and embryo of Oryzias latipes, Moina macrocopa, Simocephalus mixtus, and Pseudokirchneriella subcapitata). *Chemosphere* 75(7), 889–893. https://doi.org/10.1016/j.chemosphere.2009.01.048.
- Narayanan, M., Ali, S.S., El-Sheekh, M., 2023. A comprehensive review on the potential of microbial enzymes in multipollutant bioremediation: Mechanisms, challenges, and future prospects. *Journal of Environmental Management* 334, 117532. https://doi.org/https://doi.org/10.1016/j.jenvman.2023.117532.
- Ng, D.H., Kumar, A., Cao, B., 2016. Microorganisms meet solid minerals: Interactions and biotechnological applications. *Applied Microbiology and Biotechnology* 100(16), 6935–6946. https://doi.org/10.1007/s00253-016-7678-2.
- Noubactep, C., 2009. An analysis of the evolution of reactive species in Fe^0/H_2O systems. *Journal of Hazardous Materials* 168(2–3), 1626–1631. https://doi.org/10.1016/j.jhazmat.2009.02.143.
- Noubactep, C., Schoner, A., 2009. Fe^0-based alloys for environmental remediation: thinking

outside the box. *Journal of Hazardous Materials* 165(1-3), 1210-1214. https://doi.org/10.1016/j.jhazmat.2008.09.084.

- Okkenhaug, G., Zhu, Y.G., He, J., Li, X., Luo, L., Mulder, J., 2012. Antimony (Sb) and arsenic (As) in Sb mining impacted paddy soil from Xikuangshan, China: differences in mechanisms controlling soil sequestration and uptake in rice. *Environmental Science & Technology* 46(6), 3155-3162. https://doi.org/10.1021/es2022472.
- Okonji, S. O., Achari, G., Pernitsky, D., 2021. Environmental impacts of selenium contamination: A review on current-issues and remediation strategies in an aqueous system. *Water* 13(11), 1473-1501. https://doi.org/10.3390/w13111473.
- Olaniran, A. O., Singh, L., Kumar, A., Mokoena, P., Pillay, B., 2017. Aerobic degradation of 2, 4-dichlorophenoxyacetic acid and other chlorophenols by Pseudomonas strains indigenous to contaminated soil in South Africa: Growth kinetics and degradation pathway. *Applied Biochemistry and Microbiology* 53(2), 209-216. https://doi.org/10.1134/s0003683817020120.
- Olegario, J.T., Yee, N., Miller, M., Sczepaniak, J., Manning, B., 2009. Reduction of Se (VI) to Se(-II) by zerovalent iron nanoparticle suspensions. *Journal of Nanoparticle Research* 12(6), 2057-2068. https://doi.org/10.1007/s11051-009-9764-1.
- Oorts, K., Smolders, E., Degryse, F., Buekers, J., Gascó, G., Cornelis, G., Mertens, J., 2008. Solubility and toxicity of antimony trioxide (Sb_2O_3) in soil. *Environmental Science & Technology* 42(12), 4378-4383.
- Ouyang, H., Chen, N., Chang, G., Zhao, X., Sun, Y., Chen, S., Zhang, H., Yang, D., 2018. Selective capture of toxic selenite anions by bismuth-based metal-organic frameworks. *Angewandte Chemie* 57(40), 13197-13201. https://doi.org/10.1002/anie.201807891.
- Ouyang, T., Li, M., Appel, E., Tang, Z., Peng, S., Li, S., Zhu, Z., 2020. Magnetic response of Arsenic pollution in a slag covered soil profile close to an abandoned tungsten mine, southern China. *Scientific Reports* 10(1), 4357. https://doi.org/10.1038/s41598-020-61411-6.
- Öztürk, M., Aslan, Ş., Demirbaş, A., 2017. Accumulation of arsenic in plants from arsenic contaminated irrigation water. *Pamukkale University Journal of Engineering Sciences* 23(3), 289-297. https://doi.org/10.5505/pajes.2016.05657.
- Ozturk, Z., Tansel, B., Katsenovich, Y., Sukop, M., Laha, S., 2012. Highly organic natural media as permeable reactive barriers: TCE partitioning and anaerobic degradation profile in eucalyptus mulch and compost. *Chemosphere* 89(6), 665-671. https://doi.org/10.1016/j.chemosphere.2012.06.006.
- Pan, Y., Zhang, Y., Zhou, M., Cai, J., Li, X., Tian, Y., 2018a. Synergistic degradation of antibiotic sulfamethazine by novel pre-magnetized Fe^0/PS process enhanced by ultrasound. *Chemical Engineering Journal* 354, 777-789. https://doi.org/10.1016/j.cej.2018.08.084.

- Pan, Y., Zhang, Y., Zhou, M., Cai, J., Tian, Y., 2019a. Enhanced removal of antibiotics from secondary wastewater effluents by novel UV/pre-magnetized Fe^0/H_2O_2 process. *Water Research* 153, 144-159. https://doi.org/10.1016/j.watres.2018.12.063.
- Pan, Y., Zhang, Y., Zhou, M., Cai, J., Tian, Y., 2019b. Enhanced removal of emerging contaminants using persulfate activated by UV and pre-magnetized Fe^0. *Chemical Engineering Journal* 361, 908-918. https://doi.org/10.1016/j.cej.2018.12.135.
- Pan, Y., Zhou, M., Cai, J., Li, X., Wang, W., Li, B., Sheng, X., Tang, Z., 2018b. Significant enhancement in treatment of salty wastewater by pre-magnetization Fe^0/H_2O_2 process. *Chemical Engineering Journal* 339, 411-423. https://doi.org/10.1016/j.cej.2018.01.017.
- Pan, Y., Zhou, M., Li, X., Xu, L., Tang, Z., Liu, M., 2016. Novel Fenton-like process (pre-magnetized Fe^0/H_2O_2) for efficient degradation of organic pollutants. *Separation and Purification Technology* 169, 83-92. https://doi.org/10.1016/j.seppur.2016.06.011.
- Pan, Y., Zhou, M., Li, X., Xu, L., Tang, Z., Sheng, X., Li, B., 2017. Highly efficient persulfate oxidation process activated with pre-magnetization Fe^0. *Chemical Engineering Journal* 318, 50-56. https://doi.org/10.1016/j.cej.2016.05.001.
- Pan, Y., Zhou, M., Zhang, Y., Cai, J., Li, B., Sheng, X., 2018c. Enhanced degradation of Rhodamine B by pre-magnetized Fe^0/PS process: parameters optimization, mechanism and interferences of ions. *Separation and Purification Technology* 203, 66-74. https://doi.org/10.1016/j.seppur.2018.03.039.
- Parenky, A.C., Gevaerd de Souza, N., Asgari, P., Jeon, J., Nadagouda, M.N., Choi, H., 2020. Removal of perfluorooctanesulfonic acid in water by combining zerovalent iron particles with common oxidants. *Environmental Engineering Science* 37(7), 472-481. https://doi.org/10.1089/ees.2019.0406.
- Park, H., Kim, H., Kim, G.Y., Lee, M.Y., Kim, Y., Kang, S., 2021. Enhanced biodegradation of hydrocarbons by Pseudomonas aeruginosa-encapsulated alginate/gellan gum microbeads. *Journal of Hazardous Materials* 406, 124752. https://doi.org/10.1016/j.jhazmat.2020.124752.
- Patterson, B.M., Lee, M., Bastow, T.P., Wilson, J.T., Donn, M.J., Furness, A., Goodwin, B., Manefield, M., 2016. Concentration effects on biotic and abiotic processes in the removal of 1,1,2-trichloroethane and vinyl chloride using carbon-amended ZVI. *Journal of Contaminant Hydrology* 188, 1-11. https://doi.org/10.1016/j.jconhyd.2016.02.004.
- Pawluk, K., Fronczyk, J., Garbulewski, K., 2015. Reactivity of nano zero-valent iron in permeable reactive barriers. *Polish Journal of Chemical Technology* 17, 7-10.
- Peng, X., Wang, Z., Huang, J., Pittendrigh, B.R., Liu, S., Jia, X., Wong, P.K., 2017. Efficient degradation of tetrabromobisphenol A by synergistic integration of Fe/Ni bimetallic catalysis and microbial acclimation. *Water Research* 122, 471-480. https://doi.org/10.1016/

j.watres.2017.06.019.

- Peng, X., Wang, Z., Wei, D., Huang, Q., Jia, X., 2017. Biodegradation of tetrabromobisphenol A in the sewage sludge process. *Journal Environmental Sciences* 61, 39–48. https://doi.org/10.1016/j.jes.2017.02.023.
- Peng, X., Zhang, Z., Luo, W., Jia, X., 2013. Biodegradation of tetrabromobisphenol A by a novel Comamonas sp. strain, JXS-2-02, isolated from anaerobic sludge. *Bioresource Technology* 128, 173–179. https://doi.org/10.1016/j.biortech.2012.10.051.
- Pensini, E., Dinardo, A., Lamont, K., Longstaffe, J., Elsayed, A., Singh, A., 2019. Effect of salts and pH on the removal of perfluorooctanoic acid (PFOA) from aqueous solutions through precipitation and electroflocculation. *Canadian Journal of Civil Engineering* 46(10), 881–886. https://doi.org/10.1139/cjce-2018-0705.
- Percak-Dennett, E., He, S., Converse, B., Konishi, H., Xu, H., Corcoran, A., Noguera, D., Chan, C., Bhattacharyya, A., Borch, T., Boyd, E., Roden, E.E., 2017. Microbial acceleration of aerobic pyrite oxidation at circumneutral pH. *Geobiology* 15(5), 690–703. https://doi.org/10.1111/gbi.12241.
- Pesquer, L., Cortés, A., Pons, X., 2011. Parallel ordinary kriging interpolation incorporating automatic variogram fitting. *Computers & Geosciences* 37(4), 464–473. https://doi.org/10.1016/j.cageo.2010.10.010.
- Pezzarossa, B., Remorini, D., Gentile, M.L., Massai, R., 2012. Effects of foliar and fruit addition of sodium selenate on selenium accumulation and fruit quality. *Journal of the Science of Food and Agriculture* 92(4), 781–786. https://doi.org/10.1002/jsfa.4644.
- Phenrat, T., Thongboot, T., Lowry, G.V., 2016. Electromagnetic induction of zerovalent tron (ZVI) powder and nanoscale zerovalent iron (NZVI) particles enhances dechlorination of trichloroethylene in contaminated groundwater and soil: proof of concept. *Environmental Science & Technology* 50(2), 872–880. https://doi.org/10.1021/acs.est.5b04485.
- Phillips, D.H., 2009. Permeable reactive barriers: A sustainable technology for cleaning contaminated groundwater in developing countries. *Desalination* 248(1–3), 352–359. https://doi.org/10.1016/j.desal.2008.05.075.
- Pires, A.C., Cleary, D.F., Almeida, A., Cunha, A., Dealtry, S., Mendonca-Hagler, L.C., Smalla, K., Gomes, N.C., 2012. Denaturing gradient gel electrophoresis and barcoded pyrosequencing reveal unprecedented archaeal diversity in mangrove sediment and rhizosphere samples. *Applied and Environmental Microbiology* 78(16), 5520–5528. https://doi.org/10.1128/AEM.00386-12.
- Poothong, S., Padilla-Sanchez, J.A., Papadopoulou, E., Giovanoulis, G., Thomsen, C., Haug, L.S., 2019. Hand wipes: A useful tool for assessing human exposure to poly- and perfluoroalkyl substances (PFASs) through hand-to-mouth and dermal contacts. *Environmental Science & Technology* 53(4), 1985–1993. https://doi.org/10.1021/acs.est.8b05303.

- Preiss, A., Bauer, A., Berstermann, H.M., Gerling, S., Haas, R., Joos, A., Lehmann, A., Schmalz, L., Steinbach, K., 2009. Advanced high-performance liquid chromatography method for highly polar nitroaromatic compounds in ground water samples from ammunition waste sites. *Journal of Chromatography A* 1216(25), 4968–4975. https://doi.org/10.1016/j.chroma.2009.04.055.
- Qiu, G., Gao, T., Hong, J., Tan, W., Liu, F., Zheng, L., 2017. Mechanisms of arsenic-containing pyrite oxidation by aqueous arsenate under anoxic conditions. *Geochimica et Cosmochimica Acta* 217, 306–319. https://doi.org/10.1016/j.gca.2017.08.030.
- Qiu, H., Gui, H., Xu, H., Cui, L., Wang, C., 2022. Hydrochemical characteristics and hydraulic connection of shallow and mid-layer water in typical mining area: a case study from Sulin mining area in Northern Anhui, China. *Water Supply* 22(5), 5149–5160. https://doi.org/10.2166/ws.2022.146.
- Rajajayavel, S.R., Ghoshal, S., 2015. Enhanced reductive dechlorination of trichloroethylene by sulfidated nanoscale zerovalent iron. *Water Research* 78, 144–153. https://doi.org/10.1016/j.watres.2015.04.009.
- Ren, D., Li, S., Wu, J., Fu, L., Zhang, X., Zhang, S., 2019. Remediation of phenanthrene-contaminated soil by electrokinetics coupled with iron/carbon permeable reactive barrier. *Environmental Engineering Science* 36(9), 1224–1235. https://doi.org/10.1089/ees.2019.0066.
- Ren, J., Fan, W., Wang, X., Ma, Q., Li, X., Xu, Z., Wei, C., 2017. Influences of size-fractionated humic acids on arsenite and arsenate complexation and toxicity to Daphnia magna. *Water Research* 108, 68–77. https://doi.org/10.1016/j.watres.2016.10.052.
- Ren, M., Qu, G., Li, H., Ning, P., 2019. Influence of dissolved organic matter components on arsenate adsorption/desorption by TiO_2. *Journal of Hazardous Materials* 378, 120780. https://doi.org/10.1016/j.jhazmat.2019.120780.
- Ren, Y., Li, J., Lai, L., Lai, B., 2018. Premagnetization enhancing the reactivity of Fe^0/(passivated Fe^0) system for high concentration p-nitrophenol removal in aqueous solution. *Chemosphere* 194, 634–643. https://doi.org/10.1016/j.chemosphere.2017.12.042.
- Rubio, M. A., Lissi, E., Herrera, N., Perez, V., Fuentes, N., 2012. Phenol and nitrophenols in the air and dew waters of Santiago de Chile. *Chemosphere* 86(10), 1035–1039. https://doi.org/10.1016/j.chemosphere.2011.11.046.
- Saiers, J.E., Tao, G., 2000. Evaluation of continuous distribution models for rate-limited solute adsorption to geologic media. *Water Resources Research* 36(7), 1627–1640.
- Saigl, Z. M., 2020. Sorption behavior of selected chlorophenols onto polyurethane foam treated with iron(III): kinetics and thermodynamic study. *Environmental Monitoring and Assessment* 192(12), 748. https://doi.org/10.1007/s10661-020-08693-5.
- Saladino., R., Neri., V., Crestini., C., Costanzo., G., Graciotti., M., Mauro., E.D.,

2008. Synthesis and degradation of nucleic acid components by formamide and iron sulfur minerals. *Journal of the American Chemical Society* 130, 15512-15518.

- Samin, G., Janssen, D. B., 2012. Transformation and biodegradation of 1, 2, 3-trichloropropane (TCP). *Environmental Science and Pollution Research* 19(8), 3067-3078. https://doi.org/10.1007/s11356-012-0859-3.
- Sasaki, K., Blowes, D.W., Ptacek, C.J., Gould, W.D., 2008. Immobilization of Se(VI) in mine drainage by permeable reactive barriers: column performance. *Applied Geochemistry* 23 (5), 1012-1022. https://doi.org/10.1016/j.apgeochem.2007.08.007.
- Sathishkumar, K., Murugan, K., Benelli, G., Higuchi, A., Rajasekar, A., 2016. Bioreduction of hexavalent chromium by Pseudomonas stutzeri L1 and Acinetobacter baumannii L2. *Annals of Microbiology* 67(1), 91-98. https://doi.org/10.1007/s13213-016-1240-4.
- Schaider, L.A., Balan, S.A., Blum, A., Andrews, D.Q., Strynar, M.J., Dickinson, M.E., Lunderberg, D.M., Lang, J.R., Peaslee, G.F., 2017. Fluorinated compounds in U.S. fast food packaging. *Environmrntal Science & Technology Letter* 4(3), 105-111. https://doi.org/10.1021/acs.estlett.6b00435.
- Schnaar, G., Brusseau, M.L., 2014. Nonideal transport of contaminants in heterogeneous porous media: 11. Testing the experiment condition dependency of the continuous-distribution rate model for sorption-desorption. *Water, Air, & Soil Pollution* 225, 1-9. https://doi.org/10.1007/s11270-014-2136-1.
- Schostag, M.D., Gobbi, A., Fini, M.N., Ellegaard-Jensen, L., Aamand, J., Hansen, L.H., Muff, J., Albers, C.N., 2022. Combining reverse osmosis and microbial degradation for remediation of drinking water contaminated with recalcitrant pesticide residue. *Water Research* 216, 118352. https://doi.org/10.1016/j.watres.2022.118352.
- Sengupta, A., Jebur, M., Kamaz, M., Wickramasinghe, S.R., 2021. Removal of emerging contaminants from wastewater streams using membrane bioreactors: A review. *Membranes (Basel)* 12(1), 60-85. https://doi.org/10.3390/membranes12010060.
- Sharma, V.K., Sohn, M., McDonald, T.J., 2019. Remediation of selenium in water: a review. *Advances in Water Purification Techniques*, 203-218. https://doi.org/10.1016/b978-0-12-814790-0.00008-9.
- Shelobolina, E.S., VanPraagh, C.G., Lovley, D.R., 2003. Use of ferric and ferrous iron containing minerals for respiration by desulfitobacterium frappieri. *Geomicrobiology Journal* 20(2), 143-156. https://doi.org/10.1080/01490450303884.
- Shen, W., Wang, X., Jia, F., Tong, Z., Sun, H., Wang, X., Song, F., Ai, Z., Zhang, L., Chai, B., 2020. Amorphization enables highly efficient anaerobic thiamphenicol reduction by zero-valent iron. *Applied Catalysis B: Environmental* 264, 118550. https://doi.org/10.1016/j.apcatb.2019.118550.

- Shi, C., Cui, Y., Lu, J., Zhang, B., 2020. Sulfur-based autotrophic biosystem for efficient vanadium (V) and chromium (VI) reductions in groundwater. *Chemical Engineering Journal* 395, 124972. https://doi.org/10.1016/j.cej.2020.124972.
- Shi, G., Ma, H., Chen, Y., Liu, H., Song, G., Cai, Q., Lou, L., Rengel, Z., 2019. Low arsenate influx rate and high phosphorus concentration in wheat (Triticum aestivum L.): A mechanism for arsenate tolerance in wheat plants. *Chemosphere* 214, 94–102. https://doi.org/10.1016/j.chemosphere.2018.09.090.
- Shrimpton, H.K., Jamieson-Hanes, J.H., Ptacek, C.J., Blowes, D.W., 2018. Real-time XANES measurement of Se reduction by zerovalent iron in a flow-through cell, and accompanying Se isotope measurements. *Enviormental Science & Technology* 52(16), 9304–9310. https://doi.org/10.1021/acs.est.8b00079.
- Siggins, A., Thorn, C., Healy, M.G., Abram, F., 2021. Simultaneous adsorption and biodegradation of trichloroethylene occurs in a biochar packed column treating contaminated landfill leachate. *Journal of Hazardous Materials* 403, 123676. https://doi.org/10.1016/j.jhazmat.2020.123676.
- Silva, B., Rocha, V., Lago, A., Costa, F., Tavares, T., 2021. Rehabilitation of a complex industrial wastewater containing heavy metals and organic solvents using low cost permeable bio-barriers — From lab-scale to pilot-scale. *Separation and Purification Technology* 263, 118381. https://doi.org/10.1016/j.seppur.2021.118381.
- Silva, J. A. K., Šimůnek, J., McCray, J. E., 2020. A modified HYDRUS model for simulating PFAS transport in the vadose zone. *Water* 12(10), 2758. https://doi.org/10.3390/w12102758.
- Singh, K., Arora, S., 2011. Removal of synthetic textile dyes from wastewaters: A critical review on present treatment technologies. *Critical Reviews in Environmental Science and Technology* 41(9), 807–878. https://doi.org/10.1080/10643380903218376.
- Song, J., Wang, W., Li, R., Zhu, J., Zhang, Y., Liu, R., Rittmann, B.E., 2016. UV photolysis for enhanced phenol biodegradation in the presence of 2, 4, 6-trichlorophenol (TCP). *Biodegradation* 27(1), 59–67. https://doi.org/10.1007/s10532-016-9755-8.
- Song, J., Zhao, Q., Guo, J., Yan, N., Chen, H., Sheng, F., Lin, Y., An, D., 2019. The microbial community responsible for dechlorination and benzene ring opening during anaerobic degradation of 2,4,6-trichlorophenol. *Science of The Total Environment* 651(Pt 1), 1368–1376. https://doi.org/10.1016/j.scitotenv.2018.09.300.
- Sookhak Lari, K., Davis, G.B., Rayner, J.L., Bastow, T.P., Puzon, G.J., 2019. Natural source zone depletion of LNAPL: a critical review supporting modelling approaches. *Water Research* 157, 630–646. https://doi.org/10.1016/j.watres.2019.04.001.
- Sorensen, L., Hansen, B.H., Farkas, J., Donald, C.E., Robson, W.J., Tonkin, A., Meier, S., Rowland, S.J., 2019. Accumulation and toxicity of monoaromatic petroleum

hydrocarbons in early life stages of cod and haddock. *Environmental Pollution* 251, 212–220. https://doi.org/10.1016/j.envpol.2019.04.126.

- Souza, L.P., Graça, C.A.L., Taqueda, M.E.S., Teixeira, A.C.S.C., Chiavone-Filho, O., 2019. Insights into the reactivity of zero-valent-copper-containing materials as reducing agents of 2,4,6-trichlorophenol in a recirculating packed-column system: degradation mechanism and toxicity evaluation. *Process Safety and Environmental Protection* 127, 348–358. https://doi.org/10.1016/j.psep.2019.05.032.
- Srain, H. S., Beazley, K. F., Walker, T. R., 2021. Pharmaceuticals and personal care products and their sublethal and lethal effects in aquatic organisms. *Environmental Reviews* 29 (2), 142–181. https://doi.org/10.1139/er-2020-0054.
- Stefaniuk, M., Oleszczuk, P., Ok, Y.S., 2016. Review on nano zerovalent iron (nZVI): From synthesis to environmental applications. *Chemical Engineering Journal* 287, 618–632. https://doi.org/10.1016/j.cej.2015.11.046.
- Suanon, F., Sun, Q., Li, M., Cai, X., Zhang, Y., Yan, Y., Yu, C. P., 2017. Application of nanoscale zero valent iron and iron powder during sludge anaerobic digestion: Impact on methane yield and pharmaceutical and personal care products degradation. *Journal of Hazardous Materials* 321, 47–53. https://doi.org/10.1016/j.jhazmat.2016.08.076.
- Sun, B., Ma, J., Sedlak, D.L., 2016. Chemisorption of perfluorooctanoic acid on powdered activated carbon initiated by persulfate in aqueous solution. *Environmental Science & Technology* 50(14), 7618–7624. https://doi.org/10.1021/acs.est.6b00411.
- Sun, Y., Chen, S.S., Tsang, D.C.W., Graham, N.J.D., Ok, Y.S., Feng, Y., Li, X.D., 2017a. Zero-valent iron for the abatement of arsenate and selenate from flowback water of hydraulic fracturing. *Chemosphere* 167, 163–170. https://doi.org/10.1016/j.chemosphere.2016.09.120.
- Sun, Y., Guan, X., Wang, J., Meng, X., Xu, C., Zhou, G., 2014. Effect of weak magnetic field on arsenate and arsenite removal from water by zerovalent iron: an XAFS investigation. *Environmental Science & Technology* 48(12), 6850–6858. https://doi.org/10.1021/es5003956.
- Sun, Y., Hu, Y., Huang, T., Li, J., Qin, H., Guan, X., 2017b. Combined effect of weak magnetic fields and anions on arsenite sequestration by zerovalent iron: kinetics and mechanisms. *Environmental Science & Technology* 51(7), 3742–3750. https://doi.org/10.1021/acs.est.6b06117.
- Sun, Y., Wang, T., Peng, X., Wang, P., Lu, Y., 2016. Bacterial community compositions in sediment polluted by perfluoroalkyl acids (PFAAs) using Illumina high-throughput sequencing. *Environmental Science and Pollution Research* 23(11), 10556–10565. https://doi.org/10.1007/s11356-016-6055-0.
- Sundar, S., Chakravarty, J., 2010. Antimony toxicity. *International Journal of Environmental*

Research and Public Health 7(12), 4267–4277. https://doi.org/10.3390/ijerph7124267.

- Sunderland, E.M., Hu, X.C., Dassuncao, C., Tokranov, A.K., Wagner, C.C., Allen, J. G., 2019. A review of the pathways of human exposure to poly- and perfluoroalkyl substances (PFASs) and present understanding of health effects. *Journal of Exposure Science and Environmental Epidemiology* 29(2), 131–147. https://doi.org/10.1038/s41370-018-0094-1.
- Tao, R., Yang, H., Cui, X., Xiao, M., Gatcha-Bandjun, N., Kenmogne-Tchidjo, J.F., Lufingo, M., Konadu Amoah, B., Tepong-Tsindé, R., Ndé-Tchoupé, A. I., Touomo-Wouafo, M., Btatkeu-K, B.D., Gwenzi, W., Hu, R., Tchatchueng, J.B., Ruppert, H., Noubactep, C., 2022. The suitability of hybrid Fe^0/aggregate filtration systems for water treatment. *Water* 14(2), 260. https://doi.org/10.3390/w14020260.
- Teaf, C.M., Garber, M.M., Covert, D.J., Tuovila, B.J., 2019. Perfluorooctanoic acid (PFOA): Environmental sources, chemistry, toxicology, and potential risks. *Soil and Sediment Contamination: An International Journal* 28(3), 258–273. https://doi.org/10.1080/15320383.2018.1562420.
- Tella, M., Pokrovski, G. S., 2009. Antimony(III) complexing with O-bearing organic ligands in aqueous solution: An X-ray absorption fine structure spectroscopy and solubility study. *Geochimica et Cosmochimica Acta* 73(2), 268–290. https://doi.org/10.1016/j.gca.2008.10.014.
- Teng, X., Qi, Y., Guo, R., Zhang, S., Wei, J., Ajarem, J.S., Maodaa, S., Allam, A. A., Wang, Z., Qu, R., 2024. Enhanced electrochemical degradation of perfluorooctanoic acid by ligand-bridged Pt(II) at Pt anodes. *Journal of Hazardous Materials* 464, 133008. https://doi.org/10.1016/j.jhazmat.2023.133008.
- Tiedt, O., Mergelsberg, M., Eisenreich, W., Boll, M., 2017. Promiscuous Defluorinating enoyl-CoA hydratases/hydrolases allow for complete anaerobic degradation of 2-fluorobenzoate. *Frontiers in Microbiology* 8, 2579–2690. https://doi.org/10.3389/fmicb.2017.02579.
- Tiwari, D., Jamsheera, A., Zirlianngura, Lee, S.M., 2017. Use of hybrid materials in the trace determination of As(V) from aqueous solutions: an electrochemical study. *Environmental Engineering Research* 22(2), 186–192. https://doi.org/10.4491/eer.2016.045.
- Tran, H.L., Van, D.A., Vu, D.T., Huynh, T.H., 2022. Contamination of perfluorooctane sulfonic acid (PFOS) and perfluorooctanoic acid (PFOA) in sediment of the Cau River, Vietnam. *Environmental Monitoring and Assessment* 194(5), 380–394. https://doi.org/10.1007/s10661-022-10031-w.
- Trang., B., Li., Y., Xue., X.-S., Ateia., M., Houk., K.N., Dichtel., W.R., 2022. Low-temperature mineralization of perfluorocarboxylic acids. *Science* 377, 839–845.
- Trowbridge, J., Gerona, R.R., Lin, T., Rudel, R.A., Bessonneau, V., Buren, H., Morello-Frosch, R., 2020. Exposure to perfluoroalkyl substances in a cohort of women firefighters and office workers in San Francisco. *Environmental Science & Technology* 54(6),

3363–3374. https://doi.org/10.1021/acs.est.9b05490.

- Tzou, Y.M., Wang, S.L., Liu, J.C., Huang, Y.Y., Chen, J.H., 2008. Removal of 2,4,6-trichlorophenol from a solution by humic acids repeatedly extracted from a peat soil. *Journal of Hazardous Materials* 152(2), 812–819. https://doi.org/10.1016/j.jhazmat.2007.07.047.
- Upadhyay, S., Sinha, A., Sinha, A., 2018. Role of microorganisms in Permeable Reactive Bio-Barriers (PRBBs) for environmental clean-up: a review. *Global NEST Journal* 20(2), 269–280.
- Van Beilen, J.B., Li, Z., Duetz, W.A., Smits, T.H.M., Witholt, B., 2006. Diversity of alkane hydroxylase systems in the environment. *Oil & Gas Science and Technology* 58(4), 427–440. https://doi.org/10.2516/ogst:2003026.
- Van Genuchten, M.T., Wagenet, R.J., 1989. Two-site/two-region models for pesticide transport and degradation: Theoretical development and analytical solutions. *Soil Science Society of America Journal* 53(5), 1303–1310.
- Vignola, R., Bagatin, R., De Folly D'Auris, A., Flego, C., Nalli, M., Ghisletti, D., Millini, R., Sisto, R., 2011. Zeolites in a permeable reactive barrier (PRB): One year of field experience in a refinery groundwater — Part 1: the performances. *Chemical Engineering Journal* 178, 204–209. https://doi.org/10.1016/j.cej.2011.10.050.
- Vijayanandan, A., Philip, L., Bhallamudi, S.M., 2018. Enhanced removal of PhACs in RBF supplemented with biofilm coated adsorbent barrier: experimental and model studies. *Chemical Engineering Journal* 338, 341–357. https://doi.org/10.1016/j.cej.2017.12.099.
- Villemur, R., Lanthier, M., Beaudet, R., Lepine, F., 2006. The Desulfitobacterium genus. *Fems Microbiology Reviews* 30(5), 706–733.
- Vinceti, M., Crespi, C.M., Bonvicini, F., Malagoli, C., Ferrante, M., Marmiroli, S., Stranges, S., 2013. The need for a reassessment of the safe upper limit of selenium in drinking water. *Science of The Total Environment* 443, 633–642. https://doi.org/10.1016/j.scitotenv.2012.11.025.
- Vinceti, M., Mandrioli, J., Borella, P., Michalke, B., Tsatsakis, A., Finkelstein, Y., 2014. Selenium neurotoxicity in humans: bridging laboratory and epidemiologic studies. *Toxicology Letters* 230(2), 295–303. https://doi.org/10.1016/j.toxlet.2013.11.016.
- Vithanage, M., Rajapaksha, A.U., Dou, X., Bolan, N.S., Yang, J.E., Ok, Y.S., 2013. Surface complexation modeling and spectroscopic evidence of antimony adsorption on iron-oxide-rich red earth soils. *Journal of Colloid and Interface Science* 406, 217–224. https://doi.org/10.1016/j.jcis.2013.05.053.
- Vogel, M., Fischer, S., Maffert, A., Hubner, R., Scheinost, A.C., Franzen, C., Steudtner, R., 2018. Biotransformation and detoxification of selenite by microbial biogenesis of selenium-sulfur nanoparticles. *Journal of Hazardous Materials* 344, 749–757. https://doi.org/10.1016/j.jhazmat.2017.10.034.

- Wackett, L. P., 2022. Nothing lasts forever: understanding microbial biodegradation of polyfluorinated compounds and perfluorinated alkyl substances. *Microbial Biotechnology* 15(3), 773–792. https://doi.org/10.1111/1751-7915.13928.
- Wallace, A.R., Su, C., Sun, W., 2019. Adsorptive removal of fluoride from water using nanomaterials of ferrihydrite, apatite, and rrucite: Batch and column studies. *Environmental Engineering Science* 36(5), 634–642. https://doi.org/10.1089/ees.2018.0438.
- Wang, D., Zhang, Y., Li, J., Dahlgren, R.A., Wang, X., Huang, H., Wang, H., 2020. Risk assessment of cardiotoxicity to zebrafish (Danio rerio) by environmental exposure to triclosan and its derivatives. *Environmental Pollution* 265(Pt A), 114995. https://doi.org/10.1016/j.envpol.2020.114995.
- Wang, F., Liu, C., Shih, K., 2012. Adsorption behavior of perfluorooctanesulfonate (PFOS) and perfluorooctanoate (PFOA) on boehmite. *Chemosphere* 89(8), 1009–1014. https://doi.org/10.1016/j.chemosphere.2012.06.071.
- Wang, H., Zhong, Y., Zhu, X., Li, D., Deng, Y., Huang, W., Peng, P.a., 2021. Enhanced tetrabromobisphenol A debromination by nanoscale zero valent iron particles sulfidated with S^0 dissolved in ethanol. *Environmental Science Processes & Impacts* 23, 86. https://doi.org/10.1039/D0EM00375A.
- Wang, L., Li, Y., Zhao, Z., Cordier, T., Worms, I.A., Niu, L., Fan, C., Slaveykova, V.I., 2021. Microbial community diversity and composition in river sediments contaminated with tetrabromobisphenol A and copper. *Chemosphere* 272, 129855. https://doi.org/10.1016/j.chemosphere.2021.129855.
- Wang, M., Wang, S., Long, X., Zhuang, L., Zhao, X., Jia, Z., Zhu, G., 2019. High contribution of ammonia-oxidizing archaea (AOA) to ammonia oxidation related to a potential active AOA species in various arable land soils. *Journal of Soils and Sediments* 19, 1077–1087. https://doi.org/10.1007/s11368-018-2108-y.
- Wang, Q., Liu, M., Zhao, H., Chen, Y., Xiao, F., Chu, W., Zhao, G., 2019. Efficiently degradation of perfluorooctanoic acid in synergic electrochemical process combining cathodic electro-Fenton and anodic oxidation. *Chemical Engineering Journal* 378. https://doi.org/10.1016/j.cej.2019.122071.
- Wang, Q., Song, X., Wei, C., Jin, P., Chen, X., Tang, Z., Li, K., Ding, X., Fu, H., 2022. In situ remediation of Cr(VI) contaminated groundwater by ZVI-PRB and the corresponding indigenous microbial community responses: a field-scale study. *Science of Total Environment* 805, 150260. https://doi.org/10.1016/j.scitotenv.2021.150260.
- Wang, S., Zhang, B., Li, T., Li, Z., Fu, J., 2020. Soil vanadium(V)-reducing related bacteria drive community response to vanadium pollution from a smelting plant over multiple gradients. *Environment International* 138, 105630. https://doi.org/10.1016/j.envint.2020.105630.

- Wang, W., Cheng, Y., Kong, T., Cheng, G., 2015. Iron nanoparticles decoration onto three-dimensional graphene for rapid and efficient degradation of azo dye. *Journal of Hazardous Materials* 299, 50–58. https://doi.org/10.1016/j.jhazmat.2015.06.010.
- Wang, W., Dong, Q., Qiu, H., Li, H., Mao, Y., Liu, Y., Gong, T., Xiang, M., Huang, Y., Wang, C., Zan, R., 2022. Rapid reactivation of aged NZVI/GO by Shewanella CN32 for efficient removal of tetrabromobisphenol A and associated reaction mechanisms. *Journal of Cleaner Production* 333, 130215. https://doi.org/10.1016/j.jclepro.2021.130215.
- Wang, W., Dong, Q., Qiu, H., Li, H., Mao, Y., Liu, Y., Gong, T., Xiang, M., Huang, Y., Wang, C., Zan, R., 2022. Rapid reactivation of aged NZVIGO by Shewanella CN32 for efficient removal of tetrabromobisphenol A and associated reaction mechanisms. *Journal of Cleaner Production* 333 130215. https://doi.org/10.1016/j.jclepro.2021.130215.
- Wang, W., Gong, T., Li, H., Liu, Y., Dong, Q., Zan, R., Wu, Y., 2022. The multi-process reaction model and underlying mechanisms of 2,4,6-trichlorophenol removal in lab-scale biochar-microorganism augmented ZVI PRBs and field-scale PRBs performance. *Water Research* 217, 118422. https://doi.org/10.1016/j.watres.2022.118422.
- Wang, W., Wu, Y., 2017. Combination of zero-valent iron and anaerobic microorganisms immobilized in luffa sponge for degrading 1,1,1-trichloroethane and the relevant microbial community analysis. *Applied Microbiology and Biotechnology* 101(2), 783–796. https://doi.org/10.1007/s00253-016-7933-6.
- Wang, W., Wu, Y., 2019a. A multi-path chain kinetic reaction model to predict the evolution of 1,1,1-trichloroethane and its daughter products contaminant-plume in permeable reactive bio-barriers. *Environmental Pollution* 253, 1021–1029. https://doi.org/10.1016/j.envpol.2019.07.103.
- Wang, W., Wu, Y., 2019b. Sequential coupling of bio-augmented permeable reactive barriers for remediation of 1,1,1-trichloroethane contaminated groundwater. *Environmental Science and Pollution Research* 26(12), 12042–12054. https://doi.org/10.1007/s11356-019-04676-3.
- Wang, W., Wu, Y., Zhang, C., 2017. High-density natural luffa sponge as anaerobic microorganisms carrier for degrading 1,1,1-TCA in groundwater. *Bioprocess and Biosystems Engineering* 40(3), 383–393. https://doi.org/10.1007/s00449-016-1706-6.
- Wang, W., Zhao, L., Cao, X., 2020. The microorganism and biochar-augmented bioreactive top-layer soil for degradation removal of 2,4-dichlorophenol from surface runoff. *Science of the Total Environment* 733, 139244.
- Wang, W., Zhao, P., Hu, Y., Zan, R., 2020. Application of weak magnetic field coupling with zero-valent iron for remediation of groundwater and wastewater: a review. *Journal of Cleaner Production* 262, 121341. https://doi.org/10.1016/j.jclepro.2020.121341.
- Wang, X., He, M., Xi, J., Lu, X., 2011. Antimony distribution and mobility in rivers

around the world's largest antimony mine of Xikuangshan, Hunan Province, China. *Microchemical Journal* 97(1), 4–11. https://doi.org/10.1016/j.microc.2010.05.011.

- Wang, Y., Sun, H., Duan, X., Ang, H.M., Tadé, M.O., Wang, S., 2015. A new magnetic nano zero-valent iron encapsulated in carbon spheres for oxidative degradation of phenol. *Applied Catalysis B: Environmental* 172–173, 73–81. https://doi.org/10.1016/j.apcatb.2015.02.016.
- Wang, Z., Cousins, I. T., Berger, U., Hungerbuhler, K., Scheringer, M., 2016. Comparative assessment of the environmental hazards of and exposure to perfluoroalkyl phosphonic and phosphinic acids (PFPAs and PFPiAs): current knowledge, gaps, challenges and research needs. *Environmental International* 89–90, 235–247. https://doi.org/10.1016/j.envint.2016.01.023.
- Wang, Z., Liu, X., Ni, S.Q., Zhang, J., Zhang, X., Ahmad, H.A., Gao, B., 2017. Weak magnetic field: a powerful strategy to enhance partial nitrification. *Water Research* 120, 190–198. https://doi.org/10.1016/j.watres.2017.04.058.
- Wang., C. C., Lee., C. M., Lu., C. J., Chuang., M. S., Huang., C. Z., 2000. Biodegradation of 2,4,6-trichlorophenol in the presence of primary substrate by immobilized pure culture bacteria. *Chemosphere*, 1873–1879.
- Waskaas, M., Kharkats, Y., 1999. Magnetoconvection phenomena: A mechanism for influence of magnetic fields on electrochemical processes. *Journal of Physical Chemistry B* 103(23), 4876–4883.
- Waskaas, M., Kharkats, Y., 2001. Effect of magnetic fields on convection in solutions containing paramagnetic ions. *Journal of Electroanalytical Chemistry* 502(1), 51–57.
- Watson, C., Bahadur, K., Briess, L., Dussling, M., Kohler, F., Weinsheimer, S., Wichern, F., 2017. Mitigating negative microbial effects of *p*-nitrophenol, phenol, copper and sadmium in a sandy loam soil using biochar. *Water, Air, & Soil Pollution* 228(2). https://doi.org/10.1007/s11270-017-3243-6.
- Wee, S.Y., Aris, A.Z., 2023. Revisiting the "forever chemicals", PFOA and PFOS exposure in drinking water. Npj *Clean Water* 6(1), 57. https://doi.org/10.1038/s41545-023-00274-6.
- Wenlin.Chen, J.Wagenet, R., 1995. Solute transport in porous media with sorption-site heterogeneity. *Environmental Science & Technology* 29, 2725–2734.
- Westerhoff, P., Prapaipong, P., Shock, E., Hillaireau, A., 2008. Antimony leaching from polyethylene terephthalate (PET) plastic used for bottled drinking water. *Water Research* 42(3), 551–556. https://doi.org/10.1016/j.watres.2007.07.048.
- WHO, 2011. *Guidelines for Drinking-Waterquality*, 4th ed. 314.
- Wilopo, W., Sasaki, K., Hirajima, T., 2008a. Identification of sulfate- and arsenate-reducing bacteria in sheep manure as permeable reactive materials after arsenic immobilization

in groundwater. *Materials Transactions* 49 (10), 2275–2282. https://doi.org/10.2320/matertrans.M-MRA2008826.

- Wilopo, W., Sasaki, K., Hirajima, T., Yamanaka, T., 2008b. Immobilization of arsenic and manganese in contaminated groundwater by permeable reactive barriers using zero valent iron and sheep manure. *Materials Transactions* 49 (10), 2265–2274. https://doi.org/10.2320/matertrans.M-MRA2008827.
- Wilson, S.C., Lockwood, P.V., Ashley, P.M., Tighe, M., 2010. The chemistry and behaviour of antimony in the soil environment with comparisons to arsenic: a critical review. *Environmental Pollution* 158(5), 1169–1181. https://doi.org/10.1016/j.envpol.2009.10.045.
- Wu, F., Sun, F., Wu, S., Yan, Y., Xing, B., 2012. Removal of antimony(III) from aqueous solution by freshwater cyanobacteria Microcystis biomass. *Chemical Engineering Journal* 183, 172–179. https://doi.org/10.1016/j.cej.2011.12.050.
- Wu, J., Wang, B., Cagnetta, G., Huang, J., Wang, Y., Deng, S., Yu, G., 2020. Nanoscale zero valent iron-activated persulfate coupled with Fenton oxidation process for typical pharmaceuticals and personal care products degradation. *Separation and Purification Technology* 239, 116534. https://doi.org/10.1016/j.seppur.2020.116534.
- Wu, Y., Lin, L., Suanon, F., Hu, A., Sun, Y.-N., Yu, Z.-M., Yu, C.-P., Sun, Q., 2018. Effect of a weak magnetic field on triclosan removal using zero-valent iron under aerobic and anaerobic conditions. *Chemical Engineering Journal* 346, 24–33. https://doi.org/10.1016/j.cej.2018.03.134.
- Wu, Y., Luo, H., Wang, H., 2014. Removal of para-nitrochlorobenzene from aqueous solution on surfactant-modified nanoscale zero-valent iron/graphene nanocomposites. *Environmental Technology* 35: 21, 2698–2707. https://doi.org/10.1080/09593330.2014.919032.
- Wu, Y., Wang, C., Wang, S., An, J., Liang, D., Zhao, Q., Tian, L., Wu, Y., Wang, X., Li, N., 2021. Graphite accelerate dissimilatory iron reduction and vivianite crystal enlargement. *Water Research* 189, 116663. https://doi.org/10.1016/j.watres.2020.116663.
- Xi, J., He, M., 2013. Removal of Sb(III) and Sb(V) from aqueous media by goethite. *Water Quality Research Journal* 48(3), 223–231. https://doi.org/10.2166/wqrjc.2013.030.
- Xi, J., He, M., Wang, P., 2013. Adsorption of antimony on sediments from typical water systems in China: a comparison of Sb (III) and Sb (V) pattern. *Soil and Sediment Contamination: An International Journal* 23(1), 37–48. https://doi.org/10.1080/15320383.2013.774319.
- Xiang, M., Huang, M., Li, H., Wang, W., Huang, Y., Lu, Z., Wang, C., Si, R., Cao, W., 2021a. Nanoscale zero-valent iron/cobalt@mesoporous hydrated silica core-shell particles as a highly active heterogeneous Fenton catalyst for the degradation of tetrabromobisphenol A. *Chemical Engineering Journal* 417, 129208. https://doi.org/10.1016/j.cej.2021.129208.

- Xiang, M., Huang, M., Li, H., Wang, W., Huang, Y., Lu, Z., Wang, C., Si, R., Cao, W., 2021b. Nanoscale zero-valent iron/cobalt@mesoporous hydrated silica core-shell particles as a highly active heterogeneous Fenton catalyst for the degradation of tetrabromobisphenol A. *Chemical Engineering Journal* 417, 129208. https://doi.org/10.1016/j.cej.2021.129208.
- Xiang, S., Cheng, W., Chi, F., Nie, X., Hayat, T., Alharbi, N., 2020. Photocatalytic removal of U(VI) from wastewater via synergistic carbon-supported zero-valent iron nanoparticles and S. putrefaciens. *Applied Nano Materials* 3, 1131-1138. https://doi.org/10.1021/acsanm.9b01581.
- Xiang, W., Zhang, B., Zhou, T., Wu, X., Mao, J., 2016. An insight in magnetic field enhanced zero-valent iron/H_2O_2 Fenton-like systems: Critical role and evolution of the pristine iron oxides layer. *Scientific Reports* 6, 24094. https://doi.org/10.1038/srep24094.
- Xiao, J., Huang, J., Wang, Y., Qian, X., 2023. The fate and behavior of perfluorooctanoic acid (PFOA) in constructed wetlands: Insights into potential removal and transformation pathway. *Science of the Total Environment* 861, 160309. https://doi.org/10.1016/j.scitotenv.2022.160309.
- Xiao, X., Ulrich, B. A., Chen, B., Higgins, C. P., 2017. Sorption of poly- and perfluoroalkyl substances (PFASs) relevant to aqueous film-forming foam (AFFF)-impacted groundwater by biochars and activated carbon. *Environmental Science & Technology* 51(11), 6342-6351. https://doi.org/10.1021/acs.est.7b00970.
- Xiao, Z., Zhou, Q., Qin, H., Qiao, J., Guan, X., 2014. The enhancing effect of weak magnetic field on degradation of Orange II by zero-valent iron. *Desalination and Water Treatment* 57(4), 1659-1670. https://doi.org/10.1080/19443994.2014.974213.
- Xie, Y., Dong, H., Zeng, G., Tang, L., Jiang, Z., Zhang, C., Deng, J., Zhang, L., Zhang, Y., 2017. The interactions between nanoscale zero-valent iron and microbes in the subsurface environment: a review. *Journal of Hazardous Materials* 321, 390-407. https://doi.org/https://doi.org/10.1016/j.jhazmat.2016.09.028.
- Xin, B.P., Wu, C.H., Wu, C.H., Lin, C.W., 2013. Bioaugmented remediation of high concentration BTEX-contaminated groundwater by permeable reactive barrier with immobilized bead. *Journal of Hazardous Materials* 244-245, 765-772. https://doi.org/10.1016/j.jhazmat.2012.11.007.
- Xing, Y., Li, Q., Chen, X., Fu, X., Ji, L., Wang, J., Li, T., Zhang, Q., 2021. Different transport behaviors and mechanisms of perfluorooctanoate (PFOA) and perfluorooctane sulfonate (PFOS) in saturated porous media. *Journal of Hazardous Materials* 402, 123435. https://doi.org/10.1016/j.jhazmat.2020.123435.
- Xiong, X., Sun, B., Zhang, J., Gao, N., Shen, J., Li, J., Guan, X., 2014. Activating persulfate by Fe^0 coupling with weak magnetic field: performance and mechanism. *Water*

Research 62, 53–62. https://doi.org/10.1016/j.watres.2014.05.042.

- Xiong, X., Sun, Y., Sun, B., Song, W., Sun, J., Gao, N., Qiao, J., Guan, X., 2015. Enhancement of the advanced Fenton process by weak magnetic field for the degradation of 4-nitrophenol. *RSC Advances*, 13357–13365 https://doi.org/10.1039/C4RA16318D10.1039/c4ra16318d.
- Xiu, Z.-m., Gregory, K.B., Lowry, G.V., Alvarez, P.J.J., 2010. Effect of bare and coated nanoscale zerovalent iron on tceA and vcrA gene expression in dehalococcoides spp. *Environmental Science & Technology* 44(19), 7647–7651. https://doi.org/10.1021/es101786y.
- Xu, C., Zhang, B., Wang, Y., Shao, Q., Zhou, W., Fan, D., Bandstra, J.Z., Shi, Z., Tratnyek, P.G., 2016a. Effects of sulfidation, magnetization, and oxygenation on azo dye reduction by zerovalent iron. *Enviormental Science & Technology* 50(21), 11879–11887. https://doi.org/10.1021/acs.est.6b03184.
- Xu, C., Zhang, B., Zhu, L., Lin, S., Sun, X., Jiang, Z., Tratnyek, P.G., 2016b. Sequestration of antimonite by zerovalent iron: using weak magnetic field effects to enhance performance and characterize reaction mechanisms. *Environmental Science & Technology* 50(3), 1483–1491. https://doi.org/10.1021/acs.est.5b05360.
- Xu, H., Sun, Y., Li, J., Li, F., Guan, X., 2016. Aging of zerovalent iron in synthetic groundwater: X-ray photoelectron spectroscopy depth profiling characterization and depassivation with uniform magnetic field. *Environmental Science & Technology* 50(15), 8214–8222. https://doi.org/10.1021/acs.est.6b01763.
- Xu, J., Guo, J., Xu, M., Chen, X., 2020. Enhancement of microbial redox cycling of iron in zero-valent iron oxidation coupling with deca-brominated diphenyl ether removal. *Science of the Total Environment* 748, 141328. https://doi.org/10.1016/j.scitotenv.2020.141328.
- Xu, N., Tan, G., Wang, H., Gai, X., 2016. Effect of biochar additions to soil on nitrogen leaching, microbial biomass and bacterial community structure. *European Journal of Soil Biology* 74, 1–8. https://doi.org/10.1016/j.ejsobi.2016.02.004.
- Xu, T., Zhu, R., Shang, H., Xia, Y., Liu, X., Zhang, L., 2019. Photochemical behavior of ferrihydrite-oxalate system: Interfacial reaction mechanism and charge transfer process. *Water Research* 159, 10–19. https://doi.org/10.1016/j.watres.2019.04.055.
- Xue, A., Shen, Z.Z., Zhao, B., Zhao, H.Z., 2013. Arsenite removal from aqueous solution by a microbial fuel cell-zerovalent iron hybrid process. *Journal of Hazardous Materials* 261, 621–627. https://doi.org/10.1016/j.jhazmat.2013.07.072.
- Yang, B., Qiu, H., Zhang, P., He, E., Xia, B., Liu, Y., Zhao, L., Xu, X., Cao, X., 2022. Modeling and visualizing the transport and retention of cationic and oxyanionic metals (Cd and Cr) in saturated soil under various hydrochemical and hydrodynamic conditions. *Science of the Total Environment* 812, 151467. https://doi.org/10.1016/j.scitotenv.2021.151467.

- Yang, C., Chen, W., Chang, B., 2017. Biodegradation of tetrabromobisphenol-A in sludge with spent mushroom compost. *International Biodeterioration & Biodegradation* 119, 387–395. https://doi.org/10.1016/j.ibiod.2016.10.051.
- Yang, F., Zhang, S., Sun, Y., Cheng, K., Li, J., Tsang, D.C.W., 2018. Fabrication and characterization of hydrophilic corn stalk biochar-supported nanoscale zero-valent iron composites for efficient metal removal. *Bioresource Technology* 265, 490–497. https://doi.org/https://doi.org/10.1016/j.biortech.2018.06.029.
- Yang, H. H., Lee, K. T., Hsieh, Y. S., Luo, S. W., Huang, R. J., 2015. Emission characteristics and chemical compositions of both cilterable and condensable fine particulate from steel plants. *Aerosol and Air Quality Research* 15(4), 1672–1680. https://doi.org/10.4209/aaqr.2015.06.0398.
- Yang, J., Meng, L., Guo, L., 2018. In situ remediation of chlorinated solvent-contaminated groundwater using ZVI/organic carbon amendment in China: field pilot test and full-scale application. *Environmental Science and Pollution Research* 25(6), 5051–5062. https://doi.org/10.1007/s11356-017-9903-7.
- Yang, J., Zhou, L., Li, H., 2018. Synergistic effects of acclimated bacterial community and zero valent iron for removing 1, 1, 1-trichloroethane and 1, 4-dioxane co-contaminants in groundwater. *Journal of Chemical Technology & Biotechnology* 93(8), 2244–2251. https://doi.org/10.1002/jctb.5567.
- Yang, M., Zhang, X., Yang, Y., Liu, Q., Nghiem, L.D., Guo, W., Ngo, H.H., 2022. Effective destruction of perfluorooctanoic acid by zero-valent iron laden biochar obtained from carbothermal reduction: experimental and simulation study. *Science of the Total Environment* 805, 150326. https://doi.org/10.1016/j.scitotenv.2021.150326.
- Yang, W.J., Wu, H.B., Zhang, C., Zhong, Q., Hu, M.J., He, J.L., Li, G.A., Zhu, Z. Y., Zhu, J. L., Zhao, H. H., Zhang, H. S., Huang, F., 2021. Exposure to 2, 4-dichlorophenol, 2,4,6-trichlorophenol, pentachlorophenol and risk of thyroid cancer: a case-control study in China. *Environmental Science and Pollution Research* 28, 61329–61343. https://doi.org/10.1007/s11356-021-14898-z.
- Yang, Y., Ok, Y.S., Kim, K.H., Kwon, E.E., Tsang, Y.F., 2017. Occurrences and removal of pharmaceuticals and personal care products (PPCPs) in drinking water and water/sewage treatment plants: a review. *Science of The Total Environment* 596–597, 303–320. https://doi.org/10.1016/j.scitotenv.2017.04.102.
- Yang, Z., Wang, X., Li, H., Yang, J., Zhou, L., Liu, Y., 2017. Re-activation of aged-ZVI by iron-reducing bacterium Shewanella putrefaciens for enhanced reductive dechlorination of trichloroethylene. *Journal of Chemical and Biotechnology* 92, 2642–2649. https://doi.org/10.1002/jctb.5284.
- Yi, L., Chai, L., Xie, Y., Peng, Q., Peng, Q., 2016. Isolation, identification, and

degradation performance of a PFOA-degrading strain. *Genetics and Molecular Research* 15(2), 235–246.

- Yoon, I.-H., Kim, K.-W., Bang, S., Kim, M. G., 2011. Reduction and adsorption mechanisms of selenate by zero-valent iron and related iron corrosion. *Applied Catalysis B: Environmental* 104(1), 185–192. https://doi.org/https://doi.org/10.1016/j.apcatb.2011.02.014.
- Yu, Y., Hou, Y., Dang, Y., Zhu, X., Li, Z., Chen, H., Xiang, M., Li, Z., Hu, G., 2021. Exposure of adult zebrafish (Danio rerio) to tetrabromobisphenol A causes neurotoxicity in larval offspring, an adverse transgenerational effect. *Journal of Hazardous Materials* 414, 125408. https://doi.org/10.1016/j.jhazmat.2021.125408.
- Yu, Y., Yu, Z., Chen, H., Han, Y., Xiang, M., Chen, X., Ma, R., Wang, Z., 2019. Tetrabromobisphenol A: disposition, kinetics and toxicity in animals and humans. *Environmental Pollution* 253, 909–917. https://doi.org/10.1016/j.envpol.2019.07.067.
- Yu., F., Wang., Y., Ma., H., Zhou., M., 2020. Hydrothermal synthesis of FeS_2 as a highly efficient heterogeneous electro-Fenton catalyst to degrade diclofenac via molecular oxygen effects for Fe (II)/Fe (III) cycle. *Separation and Purification Technology* 248, 117022. https://doi.org/10.1016/j.seppur.2020.117022.
- Zaheer, M., Ullah, H., Mashwani, S.A., ul Haq, E., Ali Shah, S.H., Manzoor, F., 2021. Solute transport modelling in low-permeability homogeneous and saturated soil media. *Rudarsko Geolosko Naftni Zbornik* 36(2), 25–32. https://doi.org/10.17794/rgn.2021.2.3.
- Zazpe, R., Rodriguez Pereira, J., Thalluri, S. M., Hromadko, L., Pavlinak, D., Kolibalova, E., Kurka, M., Sopha, H., Macak, J.M., 2023. 2D FeSx nanosheets by atomic layer deposition: electrocatalytic properties for the hydrogen evolution reaction. *ChemSusChem* 16(11), 1–10. https://doi.org/10.1002/cssc.202300115.
- Zelmanov, G., Semiat, R., 2013. Selenium removal from water and its recovery using iron (Fe^{3+}) oxide/hydroxide-based nanoparticles sol (NanoFe) as an adsorbent. *Separation and Purification Technology* 103, 167–172. https://doi.org/10.1016/j.seppur.2012.10.037.
- Zhang, C., Wang, S., Lv, Z., Zhang, Y., Cao, X., Song, Z., Shao, M., 2019. $NanoFe^3O^4$ accelerates anoxic biodegradation of 3, 5, 6-trichloro-2-pyridinol. *Chemosphere* 235, 185–193. https://doi.org/10.1016/j.chemosphere.2019.06.114.
- Zhang, G., Wang, L., Xie, X., Ma, Z., Liao, Z., Qi, P., 2019. The role of scientific research in the formulation of the double-control policy for groundwater management in China. *Water Policy* 21(6), 1193–1206. https://doi.org/10.2166/wp.2019.223.
- Zhang, H., Yu, T., Li, J., Wang, Y.R., Wang, G.L., Li, F., Liu, Y., Xiong, M.H., Ma, Y.Q., 2018. Two dcm gene clusters essential for the degradation of diclofop-methyl in a microbial consortium of rhodococcus sp. JT-3 and brevundimonas sp. JT-9. *Journal of Agricultural and Food Chemistry* 66(46), 12217–12226. https://doi.org/10.1021/acs.jafc.

8b05382.

- Zhang, M., Shi, Y., Lu, Y., Johnson, A.C., Sarvajayakesavalu, S., Liu, Z., Su, C., Zhang, Y., Juergens, M.D., Jin, X., 2017. The relative risk and its distribution of endocrine disrupting chemicals, pharmaceuticals and personal care products to freshwater organisms in the Bohai Rim, China. *Science of the Total Environment* 590-591, 633-642. https://doi.org/10.1016/j.scitotenv.2017.03.011.
- Zhang, N., Gang, D.C., Lin, L.S., 2010. Adsorptive removal of parts per million level selenate using iron-coated GAC adsorbents. *Journal of Environmental Engineering* 136(10), 1089-1095 https://doi.org/10.1061//asce/ee.1943-7870.0000245.
- Zhang, Q., Wang, Y., Wang, Z., Zhang, Z., Wang, X., Yang, Z., 2021. Active biochar support nano zero-valent iron for efficient removal of U(VI) from sewage water. *Journal of Alloys and Compounds* 852, 156993. https://doi.org/10.1016/j.jallcom.2020.156993.
- Zhang, R., Leiviskä, T., Tanskanen, J., Gao, B., Yue, Q., 2019. Utilization of ferric groundwater treatment residuals for inorganic-organic hybrid biosorbent preparation and its use for vanadium removal. *Chemical Engineering Journal* 361, 680-689. https://doi.org/10.1016/j.cej.2018.12.122.
- Zhang, S., Kang, Q., Peng, H., Ding, M., Zhao, F., Zhou, Y., Dong, Z., Zhang, H., Yang, M., Tao, S., Hu, J., 2019. Relationship between perfluorooctanoate and perfluorooctane sulfonate blood concentrations in the general population and routine drinking water exposure. *Environment International* 126, 54-60. https://doi.org/10.1016/j.envint.2019.02.009.
- Zhang, X., Li, J., Sun, Y., Li, L., Pan, B., Zhang, W., Guan, X., 2018. Aging of zerovalent iron in various coexisting solutes: characteristics, reactivity toward selenite and rejuvenation by weak magnetic field. *Separation and Purification Technology* 191, 94-100. https://doi.org/10.1016/j.seppur.2017.09.020.
- Zhang, Y., Cao, B., Yin, H., Meng, L., Jin, W., Wang, F., Xu, J., Al-Tabbaa, A., 2022. Application of zeolites in permeable reactive barriers (PRBs) for in-situ groundwater remediation: a critical review. *Chemosphere* 308(Pt 1), 136290. https://doi.org/10.1016/j.chemosphere.2022.136290.
- Zhang, Y., Wang, B., Cagnetta, G., Duan, L., Yang, J., Deng, S., Huang, J., Wang, Y., Yu, G., 2018. Typical pharmaceuticals in major WWTPs in Beijing, China: Occurrence, load pattern and calculation reliability. *Water Research* 140, 291-300. https://doi.org/10.1016/j.watres.2018.04.056.
- Zhao, F. J., Ma, J. F., Meharg, A. A., McGrath, S. P., 2009. Arsenic uptake and metabolism in plants. *The New phytologist* 181(4), 777-794. https://doi.org/10.1111/j.1469-8137.2008.02716.x.
- Zhao, L., Xiao, D., Liu, Y., Xu, H., Nan, H., Li, D., Kan, Y., Cao, X., 2020.

Biochar as simultaneous shelter, adsorbent, pH buffer, and substrate of Pseudomonas citronellolis to promote biodegradation of high concentrations of phenol in wastewater. *Water Res* 172, 115494. https://doi.org/10.1016/j.watres.2020.115494.

- Zhao, Z., Pan, S., Ye, Y., Zhang, X., Pan, B., 2020. FeS2/H2O2 mediated water decontamination from p-arsanilic acid via coupling oxidation, adsorption and coagulation: Performance and mechanism. *Chemical Engineering Journal* 381. https://doi.org/10.1016/j.cej.2019.122667.
- Zheng, J., Ohata, M., Furuta, N., 2000. Studies on the speciation of inorganic antimony compounds in airborne particulate matter by HPLC-ICP-MS. *Analyst* 125, 1025–1028.
- Zhou, D., Brusseau, M.L., Zhang, Y., Li, S., Wei, W., Sun, H., Zheng, C., 2021. Simulating PFAS adsorption kinetics, adsorption isotherms, and nonideal transport in saturated soil with tempered one-sided stable density (TOSD) based models. *Journal of Hazardous Materials* 411, 125169. https://doi.org/10.1016/j.jhazmat.2021.125169.
- Zhou, S., Song, D., Gu, J.D., Yang, Y., Xu, M., 2022. Perspectives on microbial electron transfer networks for environmental biotechnology. *Frontiers in Microbiology* 13, 845796. https://doi.org/10.3389/fmicb.2022.845796.
- Zhou, T., Feng, K., Xiang, W., Lv, Y., Wu, X., Mao, J., He, C., 2018. Rapid decomposition of diclofenac in a magnetic field enhanced zero-valent iron/EDTA Fenton-like system. *Chemosphere* 193, 968–977. https://doi.org/https://doi.org/10.1016/j.chemosphere.2017.11.090.
- Zhu, B.-W., Lim, T.-T., 2007. Catalytic reduction of chlorobenzenes with Pd/Fe nanoparticles: Reactive sites, catalyst stability, particle aging, and regeneration. *Environmental Science & Technology* 41(21), 7523–7529. https://doi.org/10.1021/es0712625.
- Zhu, F., Tan, X., Zhao, W., Feng, L., He, S., Wei, L., Yang, L., Wang, K., Zhao, Q., 2022. Efficiency assessment of ZVI-based media as fillers in permeable reactive barrier for multiple heavy metal-contaminated groundwater remediation. *Journal of Hazardous Materials* 424, 127605. https://doi.org/https://doi.org/10.1016/j.jhazmat.2021.127605.
- Zhu, X., Li, W., Zhang, C., 2020. Extraction and removal of vanadium by adsorption with resin 201 * 7 from vanadium waste liquid. *Environmental Research* 180, 108865. https://doi.org/10.1016/j.envres.2019.108865.
- Zhu, X., Ni, J., 2011. The improvement of boron-doped diamond anode system in electrochemical degradation of p-nitrophenol by zero-valent iron. *Electrochimica Acta* 56(28), 10371–10377. https://doi.org/10.1016/j.electacta.2011.05.062.